Lecture Notes of the Institute for Computer Sciences, Social Informatics and Telecommunications Engineering 51

Editorial Board

Ozgur Akan
Middle East Technical University, Ankara, Turkey

Paolo Bellavista
University of Bologna, Italy

Jiannong Cao
Hong Kong Polytechnic University, Hong Kong

Falko Dressler
University of Erlangen, Germany

Domenico Ferrari
Università Cattolica Piacenza, Italy

Mario Gerla
UCLA, USA

Hisashi Kobayashi
Princeton University, USA

Sergio Palazzo
University of Catania, Italy

Sartaj Sahni
University of Florida, USA

Xuemin (Sherman) Shen
University of Waterloo, Canada

Mircea Stan
University of Virginia, USA

Jia Xiaohua
City University of Hong Kong, Hong Kong

Albert Zomaya
University of Sydney, Australia

Geoffrey Coulson
Lancaster University, UK

Joel J.P.C. Rodrigues   Liang Zhou
Min Chen   Aravind Kailas (Eds.)

Green Communications and Networking

First International Conference, GreeNets 2011
Colmar, France, October 5-7, 2011
Revised Selected Papers

 Springer

Volume Editors

Joel J.P.C. Rodrigues
University of Beira interior, Instituto de Telecomunicações
Rua Marques D'Avila e Bolama, 6201-001 Covilhã, Portugal
E-mail: joeljr@ieee.org

Liang Zhou
Nanjing University of Posts, and Telecommunications
Xinmofan Road 66, Nanjing 210003, China
E-mail: liang.zhou@ieee.org

Min Chen
Seoul National University
202, Hanul-villa, 520-9, Uijeongbu 2dong
Uijeongbu-si, Gyeonggi-do
Seoul 151-742, Korea
E-mail: minchen@ieee.org

Aravind Kailas
The University of North Carolina at Charlotte
Department of Electrical and Computer Engineering
9201 University City Bvd., Woodward Hall 230E
Charlotte 28223, NC , USA,
E-mail: aravind.kailas@uncc.edu

ISSN 1867-8211   e-ISSN 1867-822X
ISBN 978-3-642-33367-5   ISBN 978-3-642-33368-2 (eBook)
DOI 10.1007/978-3-642-33368-2

Springer Heidelberg Dordrecht London New York

Library of Congress Control Number: 2012946461

CR Subject Classification (1998): C.2.0-1, C.2.4-5, C.2.m, J.2, K.4.m

© ICST Institute for Computer Science, Social Informatics and Telecommunications Engineering 2012

This work is subject to copyright. All rights are reserved, whether the whole or part of the material is
concerned, specifically the rights of translation, reprinting, re-use of illustrations, recitation, broadcasting,
reproduction on microfilms or in any other way, and storage in data banks. Duplication of this publication
or parts thereof is permitted only under the provisions of the German Copyright Law of September 9, 1965,
in its current version, and permission for use must always be obtained from Springer. Violations are liable
to prosecution under the German Copyright Law.
The use of general descriptive names, registered names, trademarks, etc. in this publication does not imply,
even in the absence of a specific statement, that such names are exempt from the relevant protective laws
and regulations and therefore free for general use.

Typesetting: Camera-ready by author, data conversion by Scientific Publishing Services, Chennai, India

Printed on acid-free paper

Springer is part of Springer Science+Business Media (www.springer.com)

Preface

Welcome to the First ICST International Conference on Green Communications and Networking (GreeNets 2011), which was held in Colmar, France, during October 5–7, 2011. This conference aims at being the premier forum for presentation of results on cutting-edge research on green communications and networking (GCN). The mission of the conference is to share novel basic research ideas as well as experimental applications in the GCN area in addition to identifying new directions for future research and development. GreeNets 2011 provided researchers an excellent opportunity to share their perspectives with others interested in the various aspects of GCN. The conference consisted of multiple sessions that cover a broad range of research aspects. We hope that the conference proceedings will serve as a valuable reference to researchers and developers in the area.

This year, we received a large number of submissions from all over the world. All papers received rigorous peer reviews from our Technical Program Committee (TPC). After carefully examining all the received review reports, the TPC only selected very good papers for presentation at the conference and publication in this Springer LNICST book.

Putting together ICST GreeNets 2011 was a team effort. First of all, we would like to thank the authors for providing the content of the program. We would also like to express our gratitude to the TPC and reviewers, who worked very hard in reviewing papers and providing suggestions for their improvements. We would like to thank our financial sponsor ICST, technical sponsor CREATE-NET, and the technical cooperation of EAI, for their support in making GreeNets 2011 a successful event. For a list of all individuals who contributed to GreeNets 2011, please visit the conference website: http://greenets.org.

Joel J.P.C. Rodrigues
Liang Zhou

Organization

Organizing Committee

Steering Committee Chairs

Athanasios Vasilakos   National Technical University
of Athens (NTUA), Greece
(vasilako AT ath.forthnet.gr)

Imrich Chlamtac   University of Trento, Italy
(imrich.chlamtac AT create-net.org)

General Chairs

Joel Rodrigues   Institute of Telecommunications, University
of Beira Interior, Portugal
(joeljr AT ieee.org)

Liang Zhou   Nanjing University of Posts and
Telcommunications, China
(liang.zhou AT ieee.org)

Technical Program Chairs

Min Chen   Seoul National University, Korea
(minchen AT ieee.org)

Aravind Kailas   University of North Carolina Charlotte, USA
(aravindk AT ieee.org)

Publication Chair

Foad Dabiri   University of California Los Angeles, USA
(dabiri AT cs.ucla.edu)

Publicity Chairs

Binod Vaidya   University of Ottawa, Canada
(bnvaidya AT gmail.com)

Farid Farahmand   Sonoma State University, CA, USA
(farid.farahmand AT sonoma.edu)

Kai Lin   Dalian University of Technology, China
(link AT dlut.edu.cn)

Workshop Chairs

Kaushik Chowdhury   Northeastern University, USA

Hung-Yu Wei   National Taiwan University

Sponsorship and Exhibits Chair

Pascal Lorenz   University of Haute Alsace, France

Panels Chairs

Nazim Agoulmine   University of Evry, France
Hervé Guyennet   University of Besançon, France

Demos and Tutorials Chair

Nidal Nasser   University of Guelph, Canada

Poster Chair

Jaime Lloret Mauri   Polytechnic University of Valencia, Spain

Industry Chairs

Haohong Wang   TCL America, USA
Bin Wei   AT&T, USA

Special Sessions Chairs

Honggang Wang   University of Massachusetts Dartmouth, USA
Lei (Ray) Wang   Dalian University of Technology, China

Local Chair

Marc Gilg   University of Haute Alsace, France

Web Chair

Xingang Liu   Yonsei University, Korea

Technical Program Committee

Aravind Kailas   University of North Carolina Charlotte, USA
Artur Ziviani   Laboratório Nacional de Computação Científica
                                      (LNCC), Brazil
Asbjørn Hovstø   Intelligent Transportation Systems, Norway
Athanasios Vasilakos   University of Western Macedonia, Greece
Binod Vaidya   University of Ottawa, Canada
Eduardo Nakamura   FUCAPI Manaus, Brazil
Farid Farahmand   Sonoma State University, USA
Foad Dabiri   University of California at Los Angeles, USA
Hung-Yu Wei   National Taiwan University, Taiwan

Jaime Lloret Mauri	Polytechnic University of Valencia, Spain
Joel Rodrigues	Institute of Telecommunications, University of Beira Interior, Portugal
Jorge Sá Silva	University of Coimbra, Portugal
Kai Lin	Dalian University of Technology, China
Kaushik Chowdhury	Northeastern University, USA
Lei Shu	Osaka University, Japan
Liang Zhou	Nanjing University of Posts and Telcommunications, China
Min Chen	Seoul National University, Korea
Pascal Lorenz	University of Haute Alsace, France
Marc Gilg	University of Haute Alsace, France
Sherali Zeadally	University of the District of Columbia, USA
Xiaohu Ge	Huazhong University of Science and Technology, China
Yevgeni Koucheryavy	Tampere University of Technology, Finland
Costas Pattichis	University of Cyprus, Greece
Ilangko Balasingham	Rikshospitalet University Hospital, Norway
Henry Chan	The Hong Kong Polytechnic University, SAR China
Emil Jovanov	University of Alabama in Huntsville, USA
Mohammad H. Mahoor	University of Denver, USA
Liang Zhou	Hong Kong University, SAR China
Qiang Ni	Brunel University, UK
Bor-rong Chen	Harvard University , USA
Gert Cauwenberghs	University of California, USA
John Lach	University of Virginia, USA
Baozhi Chen	Rutgers University, USA
Mike Yu Chi	University of California, USA
Wei Chen	Eindhoven University of Technology, The Netherlands
Krishna Venkatasubramanian	University of Pennsylvania, USA
Kai Lin	Dalian University of Technology, China
Eryk Dutkiewicz	Macquarie University, Australia
Djamel Djenouri	CERIST Research Center, Algeria
Tony Brooks	Aalborg University Esbjerg, Denmark

Table of Contents

A New Energy Prediction Approach for Intrusion Detection in Cluster-Based Wireless Sensor Networks

Wen Shen[1,2], Guangjie Han[1,2], Lei Shu[3], Joel J.P.C Rodrigues[4],
and Naveen Chilamkurti[5]

[1] Department of Information & Communication Systems, Hohai University, China
[2] Changzhou Key Laboratory of Sensor Networks and Environmental Sensing, China
[3] Department of Multimedia Engineering, Osaka University, Japan
[4] Instituto de Telecomunicações, University of Beira Interior, Portugal
[5] Dept. of Computer Science and Computer Engineering, La Trobe University, Australia
{hanguangjie,shen.wen1986}@gmail.com, lei.shu@live.ie,
joeljr@ieee.org, n.chilamkurti@latrobe.edu.au

Abstract. Wireless Sensor Networks (WSNs) require an efficient intrusion detection scheme to identify malicious attackers. Traditional detection schemes are not well suited for WSNs due to their higher false detection rate. In this paper, we propose a novel intrusion detection scheme based on the energy prediction in cluster-based WSNs (EPIDS). The main contribution of EPIDS is to detect attackers by comparing the energy consumptions of sensor nodes. The sensor nodes with abnormal energy consumptions are identified as malicious attackers. Furthermore, EPIDS is designed to distinguish the types of denial of service (DoS) attack according to the energy consumption rate of the malicious nodes. The primary simulation experiments prove that EPIDS can detect and recognize malicious attacks effectively.

Keywords: Wireless Sensor Networks, Intrusion Detection, Energy Prediction, DoS, Attacks Recognition.

1 Introduction

Many WSNs are organized into clusters to raise their security [1]. However the broadcast nature of wireless communication causes WSNs vulnerable to various malicious attacks. More specifically these networks are vulnerable to DoS attacks due to the use of the clustering scheme in real-world scenarios. Cluster head nodes are elected to manage local clusters, which are ideal targets for adversaries to compromise. If one single node is captured by adversaries and turned into malicious head, an entire local cluster would be affected by DoS attacks. This highlights the fact that the cluster-based WSNs require an efficient intrusion detection scheme to detect DoS attacks such as selective forwarding, wormhole attack and Sybil attack etc.

There are only a few intrusion detection methods [2, 3] proposed in the research literatures which are cluster-based in WSNs. The existing intrusion detection methods can be briefly classified into two categories: signature based detection and anomaly based detection [4]. Both of these two categories focus on identifying the behaviors of

Joel J.P.C. Rodrigues et al.: (Eds.): GreeNets 2011, LNICST 51, pp. 1–12, 2012.
© Institute for Computer Sciences, Social Informatics and Telecommunications Engineering 2012

malicious nodes and consume large quantity of energy in monitoring suspicious nodes. The disadvantage of traditional intrusion detection schemes is that the network lifecycle may become shorter as the schemes process large quantity of data and transmit it frequently. Moreover the networks suffer a high false detection rate as their detection schemes are deceived by DoS attacks. Hence, traditional intrusion detection schemes are not suitable for cluster-based WSNs, and it is critical to develop an effective security mechanism for WSNs to defend DoS attacks.

In this paper, we adopt EPIDS in a cluster-based WSN. Sensor nodes can be managed locally by cluster heads. Rotating cluster heads makes it possible to elect malicious nodes as cluster heads. Adversaries can compromise any node in the network and launch DoS attacks such as selective forwarding, hello flood, wormhole, sink hole and Sybil attack. As malicious nodes require abnormal energy to launch an attack, we focus on malicious nodes' energy consumption rate in order to discover the compromised nodes.

The two notable features of our scheme are listed as follows:

> In contrary with the traditional intrusion detection methods which only detect malicious attacks based on behavior or interactions between nodes within a period of time. We believe our energy consumption rate approach in this paper is novel and has many advantages. An energy prediction method is introduced to predict all the nodes' energy consumption rate in base station and detect some energy sensitive attacks which require abnormal energy.

> Furthermore, EPIDS distinguishes various malicious attacks according to the energy consumption rate. Energy thresholds are set to classify the malicious attacks, so that we can be aware of the types of attacks.

To our best knowledge, the concept of energy prediction in intrusion detection area has never been discussed in any previous research works. These two specific features mentioned above collectively make EPIDS a new, lightweight and efficient solution that can detect various attacks applied in any cluster-based WSNs.

The rest of this paper is organized as follows: section 2 introduces related work of intrusion detection schemes in WSNs. Section 3 presents an energy prediction model of sensor nodes. Section 4 introduces the energy prediction-based intrusion detection scheme in cluster-based WSNs. In Section 5 we discuss simulation results and evaluations of our scheme.

2   Related Work

There are few existing studies in detecting and preventing DoS attacks in WSNs. Related papers [2-4] always focus on the misbehaviors of sensor nodes. The security schemes allocate considerable resources to monitor the behaviors of all the sensor nodes. After the detection of malicious nodes, most schemes establish a blacklist to isolate malicious nodes. However, none of them adopts the energy character in detecting malicious nodes.

In [4] authors proposed a technique known as Spontaneous Watchdogs. This technique use both local and global agents to watch over the communications.

For hierarchal sensor networks, global agents are activated in every cluster head. For every packet circulating in the network, global agents with the Spontaneous Watchdogs technique are able to receive both the packet and the relayed packet by the next-hop. If malicious nodes modified or selective forwarded packets, the global agents will detect the attack by Spontaneous Watchdogs.

In [5] authors proposed an insider attacker detection scheme. The scheme explores the spatial correlation existent among the networking behavior of sensors in close proximity. The author considers multiple attributes simultaneously in node behavior evaluation, with no requirement on a prior knowledge about normal or malicious sensor activities. Moreover, the scheme employs original measurements from sensors and can be employed to monitor many aspects of sensor networking behavior.

In [6] authors proposed an analytical model for intrusion detection. The authors derive the detection probability by considering two sensing models: single-sensing detection and multiple-sensing detection. In addition, the paper discusses the network connectivity and broadcast coverage, which are necessary conditions to ensure the corresponding detection probability in a WSN.

In [7] authors proposed an energy-efficient intrusion detection system for wireless sensor network based on MUSK (**M**uhammad **U**sman and **S**urraya **K**hanum) agent. The MUSK agent is installed on each node that continuously monitors intrusion. The authors assume that MUSK agents are resilient against malicious nodes that try to steal or modify information carried by the agent. However, this assumption may not be realistic in many applications.

In [8] authors proposed a group-based intrusion detection system in WSNs. The group-based intrusion detection system first divides the sensor nodes into a number of groups using δ-grouping algorithm such that the nodes in a group are physically close to each other. This feature makes it easier to detect outlier nodes and the intrusion detection results become more precise then the scheme adopts the Mahalanobis distance measurement and the OGK estimators in the intrusion detection algorithm to ensure a high breakdown point even with some missing data. However, the author assumes that there is no intense or unexpected varieties of sensed data at the grouping phase of the intrusion detection algorithm. This assumption makes the algorithm not perfect.

In [9] authors provided an energy-efficient and secure system eHIP for cluster-based WSN. The eHIP system consists of Authentication-based Intrusion Prevention (AIP) subsystem and Collaboration-based Intrusion Detection (CID) subsystem. However, collaborative monitoring of each sensor nodes would cost abundant resources of the network and low the efficiency of communication between sensor nodes.

Sensor nodes with limited resources cannot constantly monitor other nodes behavior, and report any unusual behavior to their base station or neighbor nodes. Also a compromised node can return a false alarm, which is difficult to detect. Since the nature of wireless channels implies that packet forwarding is unstable, data packets would be lost during the transmissions. Therefore, security schemes which focus on the behaviors of sensor nodes could not detect the *selective forwarding* attack efficiently. For sensor node equipped with batteries, they can not be recharged after deployment, EPIDS could analyze the energy consumption of sensor nodes and is not affected by the interference of packet loss. Our proposed approach minimizes the security control messages and eliminates the need of updating monitor reports.

3   Energy Prediction Model

We believe that malicious nodes have to use additional energy to launch DoS attacks. Therefore, we preliminarily focus on an energy prediction method to detect malicious nodes. In this paper, Markov chains model is adopted to periodically predict energy consumption of sensor nodes. The difference between the predicted and the real energy consumption of sensor nodes can be used to detect malicious nodes.

3.1   Energy Dissipation

The energy dissipation in sensor nodes depends on the energy consumption in different working states and the time they operate in each state. The sensor nodes have five operation states: 1) *Sleeping* state: a sensor node operates in *sleeping* state does not interact with other nodes. Therefore, there is no need to evaluate the trust of the sleeping node. The energy dissipation of the sleeping node in the round time is E_{sleep}; 2) *Sensing* state: in the sensor operation, sensor nodes are responsible to sensing physical parameters, such as temperature, atmospheric pressure etc.; 3) *Calculating* state: sensor nodes process the received data; 4) *Transmitting* state: sensor nodes transmit data packets between the clusters and the base station; 5) *Receiving* state: sensor nodes receive data packets.

It is believed that the energy dissipation mainly focuses on the last four states. Therefore, each sensor node can be modeled by a Markov chain [10] with the last four states.

3.2   Operation State Transition Model

As shown in Fig.1, the operation states of any sensor node shift when the node sends and receives packets, calculates data and senses information. Furthermore, the time-step is the minimum time unit of the four operation states. Each state covers several time-steps. In one time-step, state i shifts to state j with a probability of P_{ij}, for $i, j = 1, 2, 3, 4$.

In a series of n time-steps, the operation states of a sensor node can be denoted as $X = \{X_0, X_1, ..., X_n\}$. $P_{ij}^{(n)}$ represents the probability of transition from state i to state j in n time-steps. Therefore, the n-stage transition probabilities can be defined as:

$$P_{ij}^{(n)} = P\{X_n = j \mid X_1 = i\} \tag{3.1}$$

$P_{ij}^{(n)}$ can be calculated by the *Chapman-kolmogorov* equations:

$$P_{ij}^{(n)} = \sum_{k=0}^{n} P_{ik}^{(r)} P_{kj}^{(n-r)} \quad 0 < r < n \tag{3.2}$$

If a cluster head knows $P_{ij}^{(n)}$ for its sensor nodes as well as the initial states X_0 of sensor nodes, it is possible to predict the energy consumption information of all sensor nodes in the cluster. The prediction process is shown as follows: Firstly, when the sensor node is current in state i, the cluster head counts the number of time-steps the node will stay in state j, $\sum_{t=1}^{T} P_{ij}^{(t)}$ Secondly, the cluster head calculates the amount of energy dissipation in the next T time-setps, E^T :

$$E^T = \sum_{j=1}^{4} (\sum_{t=1}^{T} P_{ij}^{(t)}) * E_j \tag{3.3}$$

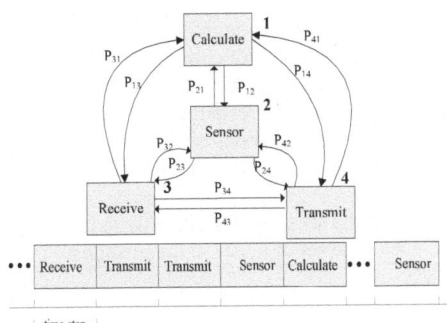

Fig. 1. The shift of operation states

Let E_j be the amount of energy dissipated in state j for one time-step. Finally, the cluster head node calculates the energy dissipation rate (ΔE) of the sensor node for the next T time-steps. The cluster head node can maintain estimations for the dissipated energy in each node by decreasing the value ΔE periodically for the amount of the remaining energy from each node. Given the energy dissipation prediction, cluster heads send the prediction results to the base station where trust information is stored.

4 Energy Prediction Based Intrusion Detection Scheme

According to the energy prediction method, EPIDS first compares the energy prediction results with the actual energy consumption at the node. Then the scheme searches nodes which spent significantly abnormal energy than other remaining nodes. The nodes with abnormal energy consumption are regarded to be malicious. Finally our scheme classifies the types of DoS attacks launched by malicious nodes.

4.1 Intrusion Detection Scheduling Algorithm

In the beginning of a round, sink node S predicts the energy consumption of each sensor node and keeps the prediction result. Then, at the end of each round, sink node is responsible for gathering energy residual of sensor nodes. Sink node broadcast an energy gathering message. On the responses to the energy gathering messages, the sensor nodes check their energy residual E_r and reply sink node with new value of E_r. If EPIDS scheme detects abnormal energy consumed at a node i, EPIDS will regard the node i as malicious and record the node's ID in a blacklist v. Sensor nodes in the blacklist will be segregated from the sensor network by removing it from the route table.

Table 1. The notations used in the intrusion detection algorithm

Notation	Meaning
S	Sink node
i	A sensor node
Round	the number of current round
E_p	the energy prediction result
E_r	the energy residual
R_{ID}	the set of nodes alive in this round
$Reply_{ID}$	the ID of node who reply this message
T	Timestamp
U	the set of nodes in the write list
v	the set of nodes in the black list

Table 2. Intrusion detection scheduling algorithm

01	If (*Not clustering*) And (receive <"*clustering*", S, *round$_i$*, u >)
02	Then
03	Broadcast <S : u, "*Energy gathering*", round, R_{ID}, T>
04	If (*node$_i$* = *alive*) and ($i \in u$) Then
05	$E_r(i)=E(i);$
06	*Round=round$_i$;*
07	Broadcast < i——> S : $E_r(i)$, round, Reply$_{ID}$, T >
08	Else if (*node$_i$* = *alive*) and ($i \in v$) Then
09	*node$_i$ is regarded as malicious and isolated;*
10	End if
11	Else if (Not clustering) And (receive < i——> S : Er(i),
12	round, Reply$_{ID}$, T >) Then
13	Store <S :i, $E_r(i)$, $E_p(i)$, round$_i$, Reply$_{ID}$,>;
14	End if
15	End if

4.2   Intrusion Detection Algorithm

The energy comparison between the energy prediction result and the energy consumption is the key to detect malicious nodes. Sink node records a set of energy residuals at the end of last round $\{r_1, r_2, r_3, ..., r_{m \times n}\}$. Then in the next round, sink node makes a prediction of energy consumption of sensor nodes, denoted as $\{p_1, p_2, p_3, ..., p_{m \times n}\}$. After receiving the residual energy $\{r_1', r_2', r_3', ..., r_{m \times n}'\}$ from all sensor nodes, the actual energy consumption is $\{r_1 - r_1', r_2 - r_2', r_3 - r_3', ..., r_{m \times n} - r_{m \times n}'\}$ calculated at the sink node. Therefore, the energy comparison of each node forms the set $\{p_1 - (r_1 - r_1'), p_2 - (r_2 - r_2'), p_3 - (r_3 - r_3'), ..., p_{m \times n} - (r_{m \times n} - r_{m \times n}')\}$. If $|p_i - (r_i - r_i')| > T_{reshold}$, $i \in [1, m \times n]$, then node i would be regarded as malicious.

4.3   Malicious Nodes Classification Algorithm

After the intrusion detection, the network identifies the types of DoS attacks launched by these malicious nodes.

Let E_c denote the energy comparison results:

$$E_c = E_p - E_r$$

E_p and E_r represent the energy prediction result and the energy real consumption of a sensor node i. k is the size of the data packet.

The possible five DoS attacks can be divided into two sets, $Attack_1$ and $Attack_2$, where $Attack_1 = \{A_1\}$ and $Attack_2 = \{A_2, A_3, A_4, A_5\}$. A_1 represents a *selective forwarding* attack; A_2, A_3, A_4, and A_5 represent *Hello flood* attack, *Sybil attack*, *Wormhole attack* and *Sinkhole attack*, respectively. $Attack_1$ is a set that include DoS attacks that energy consumptions are lower than prediction results, and $Attack_2$ is a set that includes DoS attacks that energy consumptions are lower than prediction results. To classify these five DoS attacks, our scheme sets four domains $D = \{D_1, D_2, D_3, D_4\}$ to distinguish them.

After detecting malicious nodes, EPIDS will distinguish the types of attacks. The energy comparison results not only indicate the malicious node but also lead us to the types of the attacks. Our scheme partitions the energy comparison results into four domains. The malicious nodes with the energy comparison result E_c, $E_c \in D_i$ is regarded as the node that launched with the DoS attack A_i, $i \in [1, 2, 3, 4]$.

Case 1. $E_c \geq M(E_{Tx} * k + \varepsilon_{amp} * k * d_{max}^2)$, then sensor node i is regarded as malicious one launching the *Hello flood* attack.

Case 2. $E_c \leq E_{Tx} * k + \varepsilon_{amp} * k * d_0^2$, then sensor node i is regarded as malicious one launching the *selective forwarding* attack.

Case 3. $2(E_{Tx} * k + \varepsilon_{amp} * k * d_0^2) \leq E\_c \leq (M-1)(E_{Tx} * k + \varepsilon_{amp} * k * d_0^2)$, then sensor node i is regarded as malicious one launching the *Sybil attack*.

Case 4. $(E_{Tx} * k + \varepsilon_{amp} * k * d_0^2) \leq E\_c \leq 2(E_{Tx} * k + \varepsilon_{amp} * k * d_0^2)$, then sensor node i is regarded as malicious one launching the *Wormhole attack*.

Case 5. $(M-1)(E_{Tx}*k+\varepsilon_{amp}*k*d_0^2)\leq E\_c\leq M(E_{Tx}*k+\varepsilon_{amp}*k*d_0^2)$, then sensor node i is regarded as malicious one launching the *sinkhole attack*.The printing area is 122 mm × 193 mm. The text should be justified to occupy the full line width, so that the right margin is not ragged, with words hyphenated as appropriate. Please fill pages so that the length of the text is no less than 180 mm, if possible.

5   Simulation and Performance Analysis

We use Network Simulator-2 (NS-2) to evaluate the performance of EPIDS. In order to see how the EPIDS detect the four types of DoS attacks, 100 nodes are randomly deployed in a rectangular field of size (100m×100m). Each node has an Omni-directional antenna having unity gain with a nominal radio range of 25m. The detailed parameters are shown in Table 3.

Table 3. Simulation Parameters

Parameters	Value
Number of nodes	100
Node placement	Random, uniform
Location of the Base station	50, 50
Transmission range	25m
Channel bandwidth	1Mbps
Simulation time	500 seconds
Propagation mode	Free space
Packet size	512bytes
Initial energy of each node	2J

Our security scheme detects various DoS attacks by comparing the energy consumptions and the prediction results of sensor nodes. The average energy consumption of sensor nodes along the time line is shown in Fig.2, where x-axis represents time and y-axis represents the average energy consumptions of sensor nodes.

The line with plus sign (+) represents the average energy consumption of malicious nodes launching *wormhole attacks*. As shown in Fig.2, malicious nodes spend as much as twice the energy than the prediction result in the first 60s. Then they will significantly raise the energy consumption and use up all the 2J energy at the time of 160s, while normal node just consume less than 0.5J energy. The wormhole attack can be easily detected since the wormhole attack spends nearly twice energy than the normal one. EPIDS can detect this abnormal energy consumption before 140s.

The double cross represents the average energy consumed by the nodes launching selective forwarding. The packet drop rate is set to 50%. The difference between the prediction results and the average energy consumption of *selective forwarding* attacks raises after 60s of the simulation, and EPIDS can detect this attack, when there is a difference in energy consumption larger than the preset threshold.

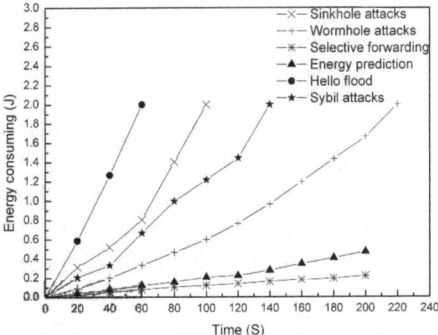

Fig. 2. Comparison of the average energy consumptions of DoS attacks and the predicted result

The star line represents the average energy consumed by the nodes launching *Sybil attacks*. The malicious node would create *M* identities with one real identity and *M-1* fake nodes. These entire *M-1* fake nodes are deployed in other clusters and would be actually controlled by the malicious node that launches the *Sybil attack*. Therefore, the malicious node would spend *M-1* times energy than the predict result. EPIDS can detect this attack when the difference in energy consumption is larger than the preset threshold.

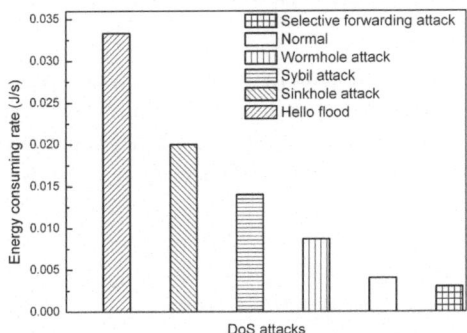

Fig. 3. The energy consumption rate of each DoS attacks

The round dotted line represents the average energy consumed by the nodes launching with *hello flood* attacks. The malicious node maximizes its broadcast range as well as the signal strength. In that case, the energy consumption would be significantly large. As can be seen in Fig.3, the nodes launching *Hello flood* can only operate 60s.

The cross line represents the average energy consumed by the nodes launching *sinkhole attacks*. The malicious nodes attract the communications of cluster heads from the other *M-1* clusters. The difference between the average energy consumption of *sinkhole attacks* and the prediction result increases gradually through the simulation. EPIDS can almost recognize *sinkhole attacks* at the beginning of the simulation. Since the energy consumption is far beyond the prediction result, *Hello flood* attack is the easiest one to be detected.

The energy consumption rate along the DoS attacks is shown in Fig.3. The *Hello flood* attack has the highest energy consumption rate 0.0333 J/s while the *selective forwarding* attack has the lowest energy consumption rate 0.00297 J/s.

Fig.4 shows the detection accuracy ratio with respect to time, where x-axis represents the time and y-axis represents the detection accuracy ratio.

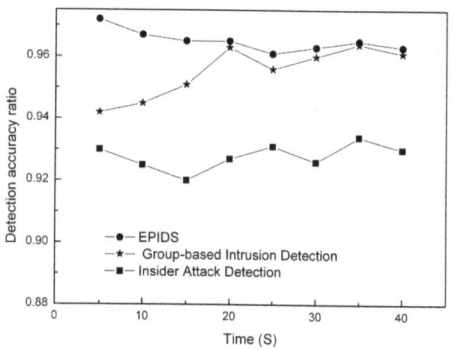

Fig. 4. The detection accuracy rate of DoS attacks

The detection accuracy of the energy prediction-based intrusion detection scheme is much higher than that of the refined group-based intrusion detection scheme. The increase in detection accuracy ratio lies in the fact that malicious nodes have to spend abnormal energy to conduct DoS attacks. This character makes these attacks easier to be detected.

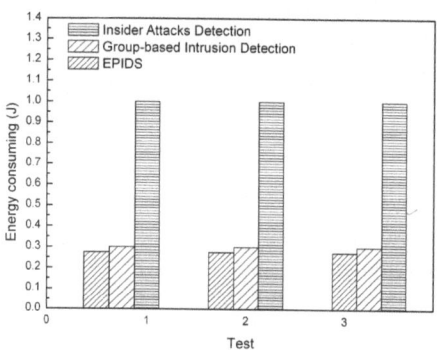

Fig. 5. Comparison of energy consumption among the refined group-based intrusion detection scheme, the insider attacker detection scheme and EPIDS.

Fig.5 shows the comparison of power consumption among the refined group-based intrusion detection scheme, the insider attacker detection scheme and EPIDS. We can see that the EPIDS consumes the least energy than that of the other two intrusion detection schemes. The reason behind this improvement lies in the fact that the energy prediction-based intrusion detection scheme does not require additional monitoring energy which is consumed in the other two schemes throughout the life time of the network.

6   Conclusions

This paper proposes a novel intrusion detection scheme for cluster-based WSNs. The proposed scheme adopts the energy prediction method to detect malicious nodes. Compared with the existing intrusion detection schemes which mainly focus on monitoring the behaviors of malicious nodes, our scheme detects malicious nodes based on the energy character. The results show that the proposed intrusion detection scheme is more efficient in detecting DoS attacks.

Acknowledgments. The work is supported by "the Excellent Master Research Funds for the Hohai University, No. XZX/09B011-02" and "the Fundamental Research Funds for the Central Universities, No.2010B22814, 2010B22914". and "the research fund of Jiangsu Key Laboratory of Power Transmission & Distribution Equipment Technology, No.2010JSSPD04". Lei Shu's research in this paper was supported by Grant-in-Aid for Scientific Research (S)(21220002) of the Ministry of Education, Culture, Sports, Science and Technology, Japan.

References

1. Wang, X., Vasilakos, A., Chen, M., Liu, Y., Kwon, T.: A Survey of Green Mobile Networks: Opportunities and Challenges. ACM/Springer Mobile Networks and Applications (2011)
2. Chen, M., Leung, V., Mao, S., Xiao, Y., Chlamtac, I.: Hybrid Geographical Routing for Flexible Energy-Delay Trade-Offs. IEEE Transactions on Vehicular Technology 58(9), 4976–4988 (2009)
3. Ssu, K.F., Wang, W.T., Chang, W.C.: Detecting Sybil attacks in Wireless Sensor Networks using neighboring information. Computer Networks 53, 3042–3056 (2009)
4. Mohi, M., Movaghar, A., Zadeh, P.M.: A Bayesian Game Approach for Preventing DoS Attacks in Wireless Sensor Networks. In: WRI International Conference on Communications and Mobile Computing, CMC 2009, vol. 3, pp. 507–511 (2009)
5. Krauß, C., Schneider, M., Eckert, C.: On handling insider attacks in wireless sensor networks. Information Security Technical Report 13(3), 165–172 (2008)
6. Wang, Y., Wang, X., Xie, B., Wang, D., Agrawal, D.P.: Intrusion Detection in Homogeneous and Heterogeneous Wireless Sensor Networks. IEEE Trans. Mobile Computing 7, 698–711 (2008)

7. Khanum, S., Usman, M., Hussain, K., Zafar, R., Sher, M.: Energy-Efficient Intrusion Detection System for Wireless Sensor Network Based on MUSK Architecture. In: Zhang, W., Chen, Z., Douglas, C.C., Tong, W. (eds.) HPCA 2009. LNCS, vol. 5938, pp. 212–217. Springer, Heidelberg (2010)
8. Li, G., He, J.A., Fu, Y.G.: Group-based intrusion detection system in wireless sensor networks. Computer Communications 31(18), 4324–4332 (2008)
9. Su, W.T., Chang, K.M., Kuo, Y.H.: eHIP: An energy-efficient hybrid intrusion prohibition system for cluster-based wireless sensor networks. Computer Networks 51(4), 1151–1168 (2007)
10. Vullers, R.J.M., Schaijk, R.V., Visser, H.J., Penders, J.H.: Energy Harvesting for Autonomous Wireless Sensor Networks. IEEE Solid-State Circuits Magazine 2(2), 29–38 (2010)

A Smart Appliance Management System with Current Clustering Algorithm in Home Network

Shih-Yeh Chen[1], Yu-Sheng Lu[2], and Chin-Feng Lai[3]

[1] Master's Program of E-Learning
Taipei Municipal University of Education
No.1, Ai-Guo West Road, Taipei, 10048 Taiwan
me_ya404@hotmail.com
[2] Business Customer Solutions Lab
Chunghwa Telecom Laboratories
No. 12, Lane 551, Min-Tsu Rd. Sec.5, Yang-Mei, Taoyuan 326, Taiwan
yusheng@cht.com.tw
[3] Institute of Computer Science and Information Engineering
National Ilan University
No. 1, Sec1, Shen-lung Road, I-Lan, 260, Taiwan
Cinfon@ieee.org

Abstract. Due to the variety of household electric devices and different power consumption habits of consumers at present, it is a challenge to identify various electric appliances without any presetting. This paper proposed the smart appliance management system for recognizing of electric appliances in home network, which can measure the power consumption of household appliances through a current sensing device. The characteristics and categories of related electric appliances are established, and this system could search the corresponding cluster data and eliminates noise for recognition functionality and error detection mechanism of electric appliances by applying the current clustering algorithm. At the same time, this system integrates household appliance control network services to control them based on users' power consumption plans, thus realizing a bidirectional monitoring service. In practical tests, the system reached a recognition rate of 95%, and could successfully control general household appliances in home network.

Keywords: Appliance Management System, Electric Appliances, Current Clustering Algorithm, Home Network.

1 Introduction

The recently promoted smart meters, such as Google PowerMeter [1] or Microsoft Hohm [2], can show the total household power consumption at present, but cannot show the power consumption of each household appliance, to say nothing of information about the household appliances that are consuming power [3-6]. As a result, users cannot further improve their power consumption habits or avoid the use

Joel J.P.C. Rodrigues et al.: (Eds.): GreeNets 2011, LNICST 51, pp. 13–24, 2012.
© Institute for Computer Sciences, Social Informatics and Telecommunications Engineering 2012

of so-called high-power electric appliances. A system that can accurately identify and detect electric appliances is a subject worthy of study. This study proposed a smart appliance management system with current clustering algorithm in home network, which can measure the household power consumption through a current sensor, transmit the data back to the energy management platform, identify each electric appliance, and then determine whether it is working normally according to its staged power consumption and various effects caused by its power sine wave intervals, so as to avoid overloading problems arising from old or faulty electrical appliances. However, the use of older or large numbers of household appliances will cause power noise problems, which could result in the inaccurate identification of electric appliances and the occurrence of errors. Therefore, in this study, a set of current clustering algorithm was presented to determine the cluster value and cluster potential for measured power information. When an abnormal value arises from the system, it is identified as noise or an abnormal state according to the clustering characteristics.

2 Smart Appliance Management System

This section introduces the overall system and expatiates on the various function modules.

2.1 Smart Meter

In this study, household power consumption was measured using a smart meter, which is mainly composed of an energy metering integrated circuit (IC), voltage and current sampling circuits, and a microprocessor, to obtain the voltage and current signals [7-14].

The energy metering IC used in this paper was the ADE7763 chip produced by Analog Devices, which can be connected with a variety of power measurement circuits, including the current converter circuit and the low resistance voltage divider circuit. During current measurement, the current analog signal is sampled from the current converter circuit, amplified through a programmable gain amplifier (PGA), and then subsequently converted to a digital signal through an analog/digital converter (ADC). Current information obtained in such a manner has large amplitude, and it must be differentiated through a high pass filter so as to obtain a correct waveform level for further integration processing through an integrator. The current signal is obtained through the root mean square (RMS) operation, and its expression is shown in Eq. 1.

$$I(RMS) = \sqrt{\frac{\int_0^T I^2(t)dt}{T}} \tag{1}$$

Due to the time signal sampling, Eq. 1 must be converted to Eq. 2.

$$I(RMS) = \sqrt{\frac{\sum_{j=1}^N I^2(j)}{N}} \tag{2}$$

The process of Eq. 2 in the hardware is as follows: after the integration of the digital signal, the square of the current signal is obtained through a multiplier, and is accumulated through a low pass filter. The RMS current value can then be obtained from the square root operation.

During the voltage measurement, the sampling method is the same as that of the current signal, but the difference is that after passing through the ADC, the analog voltage signal needs to be integrated through a low pass filter, followed by the root mean square operation. Here the instantaneous power consumption can be calculated as per Eq. 3.

$$p(t) = v(t) \times i(t) \tag{3}$$

The instantaneous voltage v(t) and instantaneous current i(t) in Eq. 3 can be expressed as Eq. 4.

$$v(t) = \sqrt{2} \times V sin(\omega t) \tag{4}$$

$$i(t) = \sqrt{2} \times I sin(\omega t) \tag{5}$$

V and I in Eqs. 4 and 5 are respectively the RMS values of the voltage and current, so Eq. 3 can be expressed as Eq. 6.

$$p(t) = VI - VI cos(2\omega t) \tag{6}$$

The actual power can be obtained through the instantaneous power, as shown in Eq. 7.

$$P = \frac{1}{nT} \times \int_0^{nT} p(t)dt = VI \tag{7}$$

Therefore, in the actual hardware implementation process, current and voltage signals are obtained through sampling and calculation, integrated through a low pass filter, and subsequently averaged after gain adjustment through a multiplier to obtain the actual power. The two parameter, I(RMS) and power (P), are applied to appliance recognition.

After the power system characteristics are measured using an energy metering IC, in order to link smart meters with the household power consumption management system, the energy metering IC performs relevant measurements after receiving commands from the microprocessor. The results are then transmitted to the microprocessor, which must have a communication interface for external transmission, allowing administrators to remotely understand the measurement results, monitor household power consumption systems, or issue additional commands to the microprocessor. The structure of the communication interface is shown in 01. After powering-on, the relays are controlled by the micro-controller unit (MCU), and sockets control whether to power on. If necessary, current values obtained from a current sensor will be converted to digital signals through a digital-to-analog converter (DAC). The digital signals are transmitted to the MCU and then to the load side. The MCU can also transmit data or commands to the central control center through ZigBee sensors [15-19], or record data in the electrically-erasable programmable read-only memory (EEPROM).

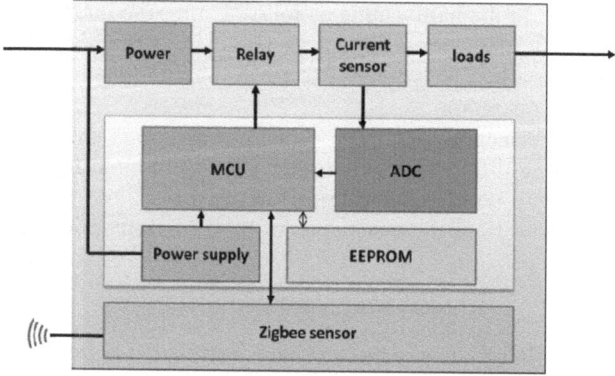

Fig. 1. The design of smart meter

2.1.1   Power Cluster

Upon receipt of the voltage and current information, the voltage information must first be normalized to 110V. Different rooms or the number of connected electric appliances are likely to affect the voltage level, which falls within a range of 110 ± 10V. During wireless transmission, the transmission interference and noise effect will often give rise to incorrect values, as shown in 0 (A), which will affect the accurate identification of electric appliances. As for an electric appliance under a variety of operating states, the conversion between its current and phase angle presents a clustering distribution, as shown in 0 (B), and the final clustering distribution will present in a fixed number of regions. According to the above characteristics, whether the current is in the same state can be judged by whether the subsequent trace value falls within the clustering range through clustering operations. If a value beyond the clustering range arises, it must be observed whether abnormal clustering values or instantaneous noise distributions arise. Through the current clustering algorithm, the recognition rate of electric appliances can be effectively improved, and abnormal error detection rates caused by noise can be reduced.

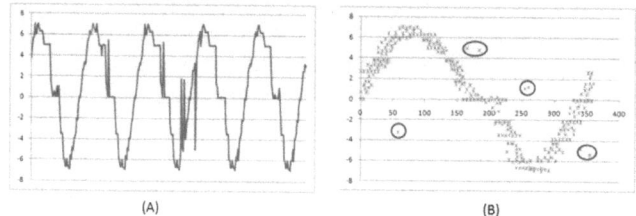

Fig. 2. The clustering distribution of current

In this study, power clustering characteristics were processed using the subtractive clustering method [20] in the neural algorithm. The main concept of the subtractive clustering method is to regard all data points as potential center points and select clustering standards according to the density of the surrounding data points. This method is independent of the complexity of the system dimension, but is proportional to the data amount. Here, it is supposed that M is the power group, and r_a is the influence distance of the center point of the clustering group and is a positive constant. The potential value P_i (Eq. 8) of the sampling point group M_i can be calculated, which represents the potential of this point becoming the clustering center point.

$$P_i = \sum_{j=1}^{n} exp\left(-\frac{\|M_i - M_j\|^2}{\frac{r_a^2}{4}}\right) \tag{8}$$

After all potential values P of the sampling points are calculated, the M_{c1} with the highest potential value is selected as the first clustering center point. The potential values of the other points then need to be modified, as per the following Eq. 9:

$$P_i = P_i - P_{c1}\, exp\left(-\frac{\|M_i - M_j\|^2}{\frac{r_b^2}{4}}\right) \tag{9}$$

wherein, r_b is a value to be set in order to avoid getting too close to the last clustering center point M_{c1}. It needs to be greater than r_a, and its value is generally recommended as 1.5 times that of r_a. After this process is repeated, the sampling point group M can be divided into subgroups, wherein, $\bar{\varepsilon}$ and $\underline{\varepsilon}$ are the upper and lower limit ratios of the potential value, which are defined in this study as 0.5 and 0.15, respectively.

2.1.2   Appliance Recognition

Each household appliance was regarded as an RLC circuit, and each had different power characteristics under different operating states. Through the previous study of the power characteristics of household appliances, they can be identified mainly based on the four parameters of I(RMS), power (P), current and voltage phase shift angle, and the distortion power factor. I(RMS) and power (P) were introduced in the last section, and here, the definitions of current and voltage phase shift angles and distortion power factor are introduced.

In an ideal AC circuit, the voltage and current should have the same phase angle, but in actual circuits, the effects of inductorsand capacitors on electric appliances have given rise to current deformations and phase difference relationships, as shown in the following 0.

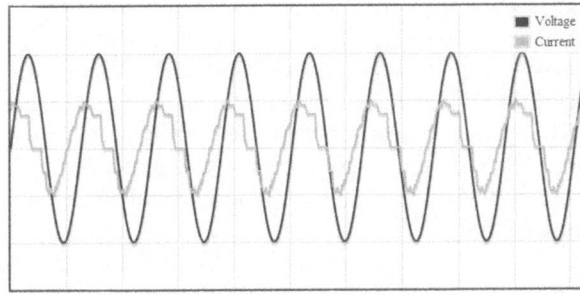

Fig. 3. The example of current deformations and phase difference relationships

The so-called distortion power factor (DPF) (Eq. 10) is defined.

$$\text{DPF} = \frac{1}{\sqrt{1+\text{THD}_i^2}} = \frac{I_{1,\text{rms}}}{I_{\text{rms}}} \tag{10}$$

THD_i is the total harmonic distortion of the load current, $I_{1,\text{rms}}$ is the fundamental component of the current, and Irms is the total current. After returning to the original waveform through the DPF, the deviation angle ω can be discovered according to the following Eq. 11:

$$|P| = |S| \cos \omega \tag{11}$$

where S is the apparent power. When the electric appliance first joins the HEM system, the system will record and study its characteristics, and then create model data. Users need to input data for different models of this appliance, such as name, brand, power usage, and so on, so as to provide enough information for the database. Later, when the electric appliance is restarted, the system will compare its characteristics with those of the model.

2.1.2.1   Comparison Programs of Power Characteristics

The system establishes a factor queue of the various eigenvalues in sequence. After the factor queue of eigenvalues is ready, when data from new electric appliances are generated, they are input into the search system. Searched results are obtained and saved as a database. Each element within it is built with the same structure, which records the device model, device importance, video description, and power characteristics. The structure of the power characteristics is comprised of four parameters, namely the power cluster value, the angular position of the peak value of the sine wave interval, the delayed value of the sine wave interval, and the displacement value of the state power. Afterwards the retriever performs the corresponding operation of the database from the factor queue based on the factor properties. Cases of factor operations are as follows.

The Operation of power characteristics: The same power characteristics of the database are compared to eliminate elements with overlarge differences in the power characteristics of the database. Here, a set of appliance-matching algorithms is presented, which is a modified algorithm based on the Boyer-Moore algorithm. It is assumed that the system regularly captures the clustering power of every section of the electric appliance in both the first D1 seconds and the last D1 seconds as the identification standard, and that the system will capture the currently measured power clustering in the first D1 seconds as the eigenvalue of power clustering for the first identification. The system compares each phase of the power clustering in the first D1 seconds with the power clustering list obtained through the first phase of screening as a comparison target. If M1 videos are successfully identified as being related, then the following electric quantity clustering data are identified with the M1 elements in the last D1 seconds as the samples and the rest may be deduced by analogy, until a complete power model is identified and the identification algorithm is completed, as shown in 0.

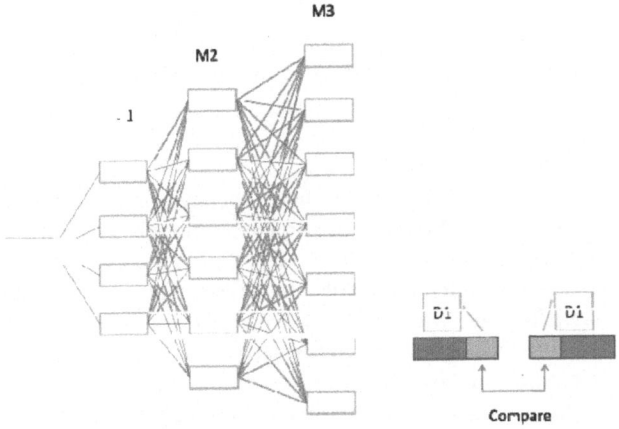

Fig. 4. Comparison programs of power characteristics

Suppose that there are M electric appliance models. It takes $2M * T_1(D1)$ to find the electric power models with a beginning and ending of D1, respectively, and $T_2(D1)$ to identify and compare the D1. The total time consumption is as Eq. 12:

$$T(M) = 2M * T1(D1) + M1 * T2(D1) + M1M2 * T2(D1) \\ + M1M2M3 * T2(D1) + \cdots \qquad (12)$$

By analogy, suppose that the same amount of models is successfully identified each time, and that $M1 = M2 ... = n$, it is simplified as Eq. 13:

$$T(M) = 2M * T1(D1) + nT2(D1) * (nm - 1)/(n - 1) \qquad (13)$$

When the operation corresponding to each factor is completed, an index array will be obtained. The index array records the element index in the database and continuously inputs the structural sequence of this device into the operating queue in sequence. The system shows the operating queue as the electric appliance being used, and the continuous control commands may also collect device models from this operating queue.

2.1.3   Context Aware Service
This service detects the regional context, such as temperature and humidity, whether there are human activities or not, and so on, and then transmits the results to the server through a network so as to be available for relevant people and other programs. It is mainly divided into four steps.

2.1.3.1   *Information Collection*
When users interact with the device, data will be collected. The system submits the collected amount of environmental parameters and related information of the interactive devices (such as UPnP, Bluetooth, infrared, ZigBee, etc.) to the learning system for information analysis, so as to produce effective information. Furthermore, useful information can be integrated with other information once again to form new

information. The detailed degree of users' behavioral patterns expressed by the information depends on the integration degree; that is, users' ongoing behaviors can be more accurately described according to the information with a higher integration degree.

2.1.3.2   Information Analysis

Information analysis aims to determine users' behavioral patterns represented by the information. Upon receipt of the information and after analysis, the system will divide interactive information into two groups. The first group is information on further behavioral patterns after interaction with the device. The second group is information on user dissatisfaction with the environment or the response time after implementation of the operation. Information analysis integrates the synthetic information in the information collection stage, and then sorts out and strengthens the programs that can be judged by the command reasoning system.

3   Result and Analysis

This section introduces the results achieved in this study and experimental analysis according to this system.

3.1   Implementation

The result interface achieved in this study is shown in the following 0. This study was mainly used to measure the power consumption of general household appliances and to identify electric appliances, as well as allow users to remotely inquire about related information through the Internet. The measured information was comprised of the household temperature and humidity measured by sensors in the surrounding environment, which could simultaneously display the power, voltage and current information. The general relay controls and IR controls of a television and an electric fan were completed in the electric control section. The control method section was

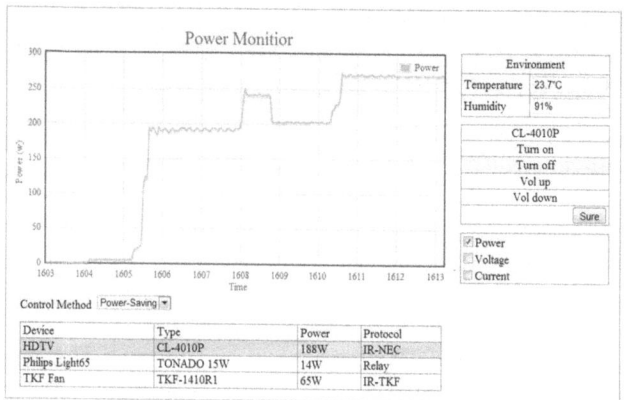

Fig. 5. The user interface of the system

divided into user-controls, power-saving, content-awareness and planning. User-control was defined as all electric appliances that are self-controlled by users and power-saving was defined as an automatic shutoff to the power supply when a standby state was detected. Content-awareness was defined as the automation of controls in electric appliances based on environmental information and historical user information. The planning section was defined as the user input of scheduled power consumption, and then allowing the system to decide the schedule the electric appliances according to the electricity price and necessity.

3.2   Experiments and Analysis

In this study, a total of 40 different household appliances were used for experimental analysis. During the experiment, at most six electric appliances were randomly started for identification analysis, and there were 30 experiments in each stage.

3.2.1   Relation between Recognition Accuracy and Recognition Time

In this experiment, the recognition accuracy was studied mainly based on the sampling time and recognition time, and the accuracy rate was defined as:

$$F = \frac{\text{Recognition time}}{\text{Sampling time}} * 100\% \tag{22}$$

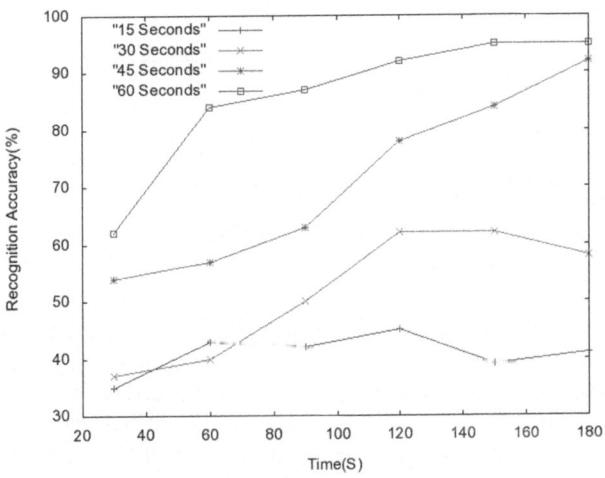

Fig. 6. Relation between recognition accuracy and recognition time

As can be seen from the experimental results, **0**, when the sampling time was insufficient (15 S), the sample would be incompletely established, resulting in recognition difficulties. Even if the recognition time was extended, it was still difficult to increase the recognition rate, while too short of recognition time would also cause recognition difficulties. As can be seen from the experimental results, the best sampling time was 60 seconds. When the recognition time reached 120 seconds, the recognition accuracy was 92%, and this could reach as high as 95%.

3.2.2   Relation between Recognition Accuracy and the Current Clustering Algorithm

In this experiment, 0, the effect of the current clustering algorithm on the system recognition rate was analyzed. An experiment is performed with a sampling time of 60 seconds and a recognition time of 90, 120, 150, and 180 seconds, respectively. The experimental results are shown in the following figure. As can be seen from the experimental results, the recognition accuracy rate was approximately 91.5% on average using the current clustering algorithm, and 79% on average without the current clustering algorithm, due to the effect of noise on data collection, sampling and identification.

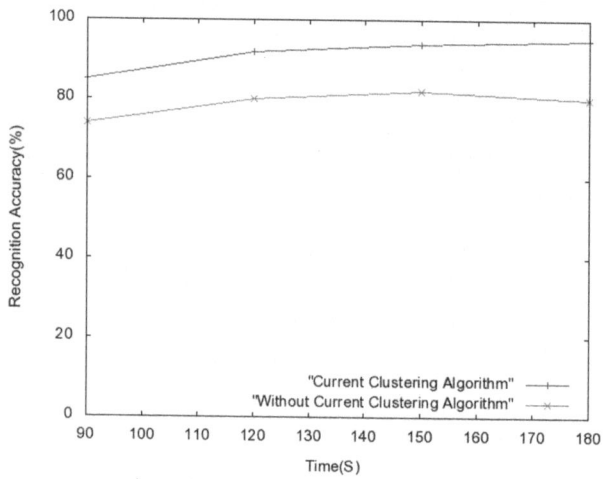

Fig. 7. Relation between recognition accuracy and the current clustering algorithm

4   Concluding Remarks

In this study, a smart appliance management system with current clustering algorithm in home network was presented. It measured power information through a smart meter and transmitted the data back to the management platform via wireless transmission. It allowed users to realize the currently used electric appliances and their power consumptions through the identification of the devices, and provided corresponding control interfaces for the users to remotely control household appliances. It established content-aware service functions with the help of context information sensors and user habits, and its recognition rate reached as high as 95% with the aid of the current clustering algorithm and the establishment of identification samples. In the future, research will be mainly engaged in establishing the planning control model matched with cloud services, so as to expand the scope of recognition and the obtainment of identification samples.

References

[1] Google Inc., Save energy. Save money. Make a difference (March 2011)

[2] Microsoft Corp., How energy efficient is your home? (March 2011)

[3] Ye, Y., Li, B., Gao, J., Sun, Y.: A design of smart energy-saving power module. In: Proc. of the 2010 5th IEEE Conference on Industrial Electronics and Applications, Taichung, pp. 898–902 (June 2010)

[4] Serra, H., Correia, J., Gano, A.J., de Campos, A.M., Teixeira, I.: Domestic power consumption measurement and automatic home appliance detection. In: Proc. of International Workshop on Intelligent Signal Processing, Faro, Portugal, pp. 128–132 (September 2005)

[5] Cho, H.S., Yamazaki, T., Hahn, M.: Determining location of appliances from multi-hop tree structures of power strip type smart meters. IEEE Transactions on Consumer Electronics 55(4), 2314–2322 (2009)

[6] Tajika, Y., Saito, T., Teramoto, K., Oosaka, N., Isshiki, M.: Networked home appliance system using Bluetooth technology integrating appliance control/monitoring with Internet service. IEEE Transactions on Consumer Electronics 49(4), 1043–1048 (2003)

[7] Han, D.M., Lim, J.H.: Design and implementation of smart home energy management systems based on ZigBee. IEEE Transactions on Consumer Electronics 56(3), 1417–1425 (2010)

[8] Han, D.M., Lim, J.H.: Smart home energy management system using IEEE 802.15.4 and ZigBee. IEEE Transactions on Consumer Electronics 56(3), 1403–1410 (2010)

[9] Bennett, C., Highfill, D.: Networking AMI Smart Meters. In: Proc. of IEEE Energy 2030 Conference, Atlanta, GA, pp. 1–8 (November 2008)

[10] Liu, J., Zhao, B., Wang, J., Zhu, Y., Hu, J.: Application of power line communication in smart power Consumption. In: Proc. of IEEE International Symposium on Power Line Communications and Its Applications, Rio de Janeiro, pp. 303–307 (March 2010)

[11] Son, Y.S., Pulkkinen, T., Moon, K.D., Kim, C.: Home energy management system based on power line communication. IEEE Transactions on Consumer Electronics 56(3), 1380–1386 (2010)

[12] Lien, C.H., Bai, Y.W., Chen, H.C., Hung, C.H.: Home appliance energy monitoring and controlling based on Power Line Communication. In: Proc. of Digest of Technical Papers International Conference on Consumer Electronics, Las Vegas, NV, pp. 1–2 (January 2009)

[13] Jahn, M., Jentsch, M., Prause, C.R., Pramudianto, F., Al-Akkad, A., Reiners, R.: The Energy Aware Smart Home. In: Proc. of 5th International Conference on Future Information Technology, Busan, pp. 1–8 (May 2010)

[14] Capone, A., Barros, M., Hrasnica, H., Tompros, S.: A New Architecture for Reduction of Energy Consumption of Home Appliances. In: TOWARDS eENVIRONMENT, European Conference of the Czech Presidency of the Council of the EU (2009)

[15] Heo, J., Hong, C.S., Kang, S.B., Jeon, S.S.: Design and Implementation of Control Mechanism for Standby Power Reduction. IEEE Transactions on Consumer Electronics 54(1), 179–185 (2008)

[16] Sundramoorthy, V., Liu, Q., Cooper, G., Linge, N., Cooper, J.: DEHEMS: A user-driven domestic energy monitoring system. In: Proc. of Internet of Things, Tokyo, November 29-December 1, pp. 1–8 (2010)

[17] Park, S., Kim, H., Moon, H., Heo, J., Yoon, S.: Concurrent simulation platform for energy-aware smart metering systems. IEEE Transactions on Consumer Electronics 56(3), 1918–1926 (2010)

[18] Park, W.K., Han, I., Park, K.R.: ZigBee based Dynamic Control Scheme for Multiple Legacy IR Controllable Digital Consumer Devices. IEEE Transactions on Consumer Electronics 53, 172–177 (2007)

[19] Min, C., Gonzalez, S., Leung, V., Qian, Z., Ming, L.: A 2G-RFID-based e-healthcare system. IEEE Wireless Communications 17, 37–43 (2010)

[20] Chiu, S.L.: Fuzzy model identification based on cluster estimation. Journal of Intelligent and Fuzzy Systems 2, 267–278 (1994)

A Flexible Boundary Sensing Model
for Group Target Tracking in Wireless Sensor Networks

Quanlong Li, Zhijia Zhao, Xiaofei Xu, and Qingjun Yan

School of Computer Science and Technology, Harbin Institute of Technology,
Harbin, P.R. China, 150001
liquanlong@hit.edu.cn

Abstract. Group target usually covers a large area and is more difficult to track in wireless sensor networks. In traditional methods, much more sensors are activated and involved in tracking, which causes a heavy network burden and huge energy cost. This paper presents a Boundary Sensing Model (BSM) used to discover group target's contour, which conserves energy by letting only a small number of sensors – BOUNDARY sensors participate in tracking. Unlike previous works, the proposed BSM is flexible by adjusting the boundary thickness thresholds. We analytically evaluate the probability of becoming a BOUNDARY sensor and the average quantity of BOUNDARY sensors, which proved to be affected by communication radius, density, and boundary thickness thresholds. Extensive simulation results confirm that our theoretical results are reasonable, and show that our proposed BSM based group target tracking method uses less number of sensors for group tracking without precision loss.

Keywords: Sensor Networks, Sensing Model, Boundary Sensing Model, Group Target Tracking.

1 Introduction

A group target is a set of individual targets moving coherently.As the targets move closely with each other, it is unpractical or unnecessary to localize every specific target in wireless sensor networks, especially when the scale of the group is relatively large (e.g. motorcade, tank column, or a herd of buffalos). In traditional methods, all discovering sensors are involved in tracking. With group target's scale increasing, the network burden and energy cost will be rather considerable. In this paper, we devise a flexible boundary sensing model to discover the group target's contour, in which only a small part of discovering sensors are involved in tracking and the tracking sensors' quantity is adjustable by some customized parameters.

The target tracking problem in WSNs has been a topic of extensive study under different metrics and assumptions [2-4,7]. However, most of them focused on individual target tracking [1-3].For group target tracking [6], some irradiative ideas of tracking targets through boundary detection are proposed in prior work [1, 5, 8]. The authors in [8] consider targets and events of interest are presented in a region, and

Joel J.P.C. Rodrigues et al.: (Eds.): GreeNets 2011, LNICST 51, pp. 25–36, 2012.
© Institute for Computer Sciences, Social Informatics and Telecommunications Engineering 2012

devise a region-based evolving targets tracking algorithm. A problem of contour tracking had been studied in [1], in which the boundaries of blobs of interest were tracked and topological changes were captured.

In this paper, we make the following contribution. We map the group target boundary sensing problem to the test of whether a sensor's Discovering Neighbors Ratio is within its thresholds. Based on our mapping, we use tool of Geometric Probability to analytically evaluate the probability of a sensor becomes a boundary sensor and the average number of boundary sensors.

Our formulations show that the boundary thickness is independent of the sensor's sense radius, and depends on the sensors' density and communication radius. Further, given a sensor deployed in a monitoring area, its probability of becoming a boundary sensor only depends on the boundary thickness thresholds. Based on these analysis results, the formulation to calculate the average number of boundary sensors is given.

The rest of this paper is organized as follows: Section 2 describes the setup of Boundary Sensing Model. In section 3, we analyze BSM theoretically. And some useful formulations are also given in this section. Section 4 states a group target tracking scheme based on BSM. The validation of our theorectical results and simulation of group target tracking are shown in Section 5. We summarize our work in Section 6.

2   Boundary Sensing Model (BSM) Setup

In [1], the authors designed a simple method to detect continuous object's boundary, but the thickness of boundary is fixed. In this section, a more flexible sensing model called Boundary Sensing Model (BSM) is proposed, where the thickness of boundary is adjustable.

The model is built upon the binary sensing model [2,3], which is famous for its minimal requirement about sensing capabilities and ease to extend other kinds of sensors. In binary sensing model, sensors are placed in two categories: discovering sensors (output 1) and non-discovering sensors (output 0). We denote this character as *discovering status (DS)*.

Definition 1. *Discovering neighbors ratio (DNR). Given sensors set I and sensor i, let* $\Theta(i)$ *be i's neighbors set and* $\Phi(i)$ *be a set in which the elements are i's neighbors discovering target (named discovering neighbors). Then i's discovering neighbors ratio is defined as*

$$\eta : I \rightarrow [0,1]$$

$$\eta(i) = \frac{|\Phi(i)|}{|\Theta(i)|}, \text{ for all } i \in I \tag{1}$$

where $|A|$ *denotes the elements number in set A .*

DNR describes the ratio of a sensor's discovering neighbors to its total ones. There is a *neighbors table* in each sensor, where its neighbors' DS is stored, as shown in Table 1. Through looking up the table, DNR can be calculated.

Table 1. Neighbors table

Neighbors' ID	DS
25	0
27	1
34	1
...	...
23	0

According to formula (1), *boundary status* can be defined as:

Definition 2. *Boundary status (BS). Given sensor $i, i \in I$, i's boundary status is defined as:*

$$\chi : I \rightarrow \{INNER, \ BOUNDARY, \ OUTER\},$$

$$\chi(i) = \begin{cases} OUTER, & \eta(i) \leq H_0, \\ BOUNDARY, & H_0 < \eta(i) < H_1, \ \text{for all } i \in I. \\ INNER, & \eta(i) \geq H_1. \end{cases} \quad (2)$$

where H_0 and H_1, $0 \leq H_0 < H_1 \leq 1$ are parameters to adjust the thickness of boundary. The BS describes whether one sensor is on the boundary of a region which contains a group target. When $H_0 = 0$ and $H_1 = 1$, the set of all BOUNDARY sensors is named Max BOUNDARY Sensors Set, denoted as $\mathcal{B}$, $\mathcal{B} \subseteq I$; otherwise, it's named as Adjustable BOUNDARY Sensors Set, denoted as B, $B \subseteq \mathcal{B} \subseteq I$.

Definition 3. *Boundary sensing model (BSM): The boundary sensing model is a sensing model, in which every sensor can compute its BS by communicating with its neighbors.*

Figure 1 illustrates a scene of a group target appearing in a region deployed with numerous BSM sensors.

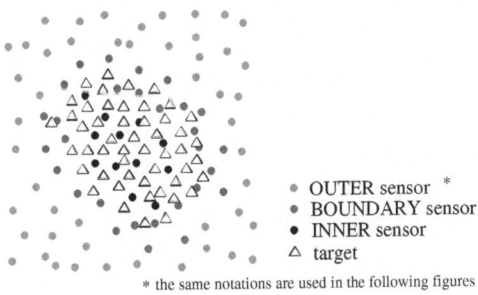

●	OUTER sensor *
●	BOUNDARY sensor
●	INNER sensor
△	target

* the same notations are used in the following figures

Fig. 1. A group target in BSM sensor networks

Every sensor is in the OUTER status initially. Once the signal strength of events it captured exceeds a certain threshold, the sensor turns into discovering status. Meanwhile, it also has a responsibility of notifying its neighbors to update their neighbors tables. Based on the updated neighbors table, a sensor could calculate its DNR and decide which BS it will become. There are three different of situations: i) if DNR≤H0, it gets into the OUTER status; ii) if DNR≥H1, it gets into the INNER status; iii) otherwise, it gets into the BOUNDARY status. The whole process is completely distributed, and only needs local communication among sensors. In Figure 2, boundary status transition diagram is shown. We assume there isn't any transition between INNER and OUTER. As the target moves continuously, it is feasible by adjusting sensor networks' parameters, such as sensors density.

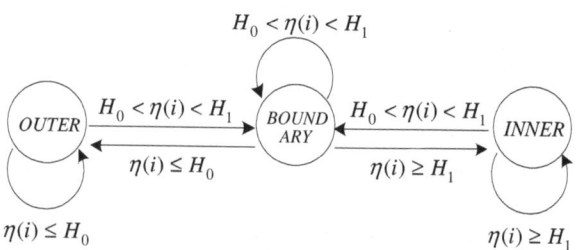

Fig. 2. Boundary status transition diagram

3   Analysis of BSM

We assume that N sensors are identically and independently distributed within a planar area, according to a random (uniform) distribution with the density of ρ. And every sensor has the same communication radius R_{comm} and the same sense radius R_{sense}. We define d as the average distance between two sensors. To ensure a sensor can communicate with its neighbors, we assume that $R_{comm} > d$.

First, we analyze boundary thickness using the number of sensors as it's metric. We define the sense line as a vertical line with a distance of R_{sense} to the targets area line, as illustrated in Figure 3.

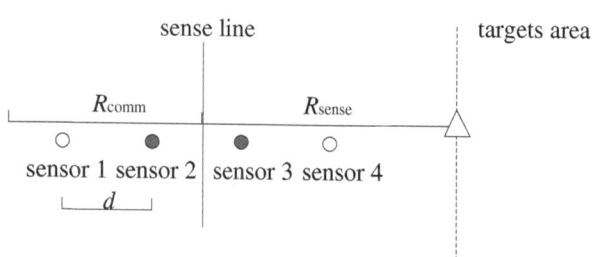

Fig. 3. A scheme for Boundary thickness analysis

Theorem 1: *Given a Max BOUNDARY Sensors Set $\mathcal{B}$, sensor communication radius R_{sense} and sensors density ρ, the boundary thickness of $\mathcal{B}$ is given by*

$$T_{\text{Boundary}} = 2 \times \left\lfloor \frac{R\text{comm}}{\sqrt{1/\rho}} \right\rfloor, \quad R\text{comm} > \sqrt{1/\rho} \tag{3}$$

Proof: *Since $R_{comm} > d$, the two closest sensors on either side of the sense line must be BOUNDARY sensors, as the sensor 2 and sensor 3 shown in Figure 3. Further, we can find such a rule.*

$$T_{\text{Boundary}} = \begin{cases} 2, & d \leq R\text{comm} < 2d \\ 4, & 2d \leq R\text{comm} < 3d \\ 6, & 3d \leq R\text{comm} < 4d \\ \cdots \end{cases} \tag{4}$$

from which we can conclude that

$$T_{\text{Boundary}} = 2 \times \left\lfloor \frac{R\text{comm}}{d} \right\rfloor = 2 \times \left\lfloor \frac{R\text{comm}}{\sqrt{1/\rho}} \right\rfloor, \quad R\text{comm} > \sqrt{1/\rho} \tag{5}$$

In Figure 3, we note that the sense line's location is used to analyze boundary thickness, without considering of R_{sense}.

Lemma 1: *Given sensor i, $i \in \mathcal{B}$, let $\eta(i)$ be sensor i's discovering neighbors ratio. Then we must have*

$$\eta(i), \ \eta(i) \in Y = \{\frac{1}{|\Theta(i)|}, \frac{2}{|\Theta(i)|}, \cdots, \frac{|\Theta(i)|-1}{|\Theta(i)|}\} \tag{6}$$

is a discrete random variable, and the probability that it takes on each value is given by

$$\Pr(\eta(i) = \frac{1}{|\Theta(i)|}) = \Pr(\eta(i) = \frac{2}{|\Theta(i)|}) = \cdots$$
$$= \Pr(\eta(i) = \frac{|\Theta(i)|-1}{|\Theta(i)|}) = \frac{1}{|\Theta(i)|-1} \tag{7}$$

Proof: *Since sensor i is uniformly distributed over interval $[a, b]$, as shown in Figure 4. It follows that*

$$\Pr(i \text{ in } [a, c]) = \Pr(i \text{ in } [c, b]) = 0.5$$

Then $\forall j \in \Theta(i)$, we must have

$$\Pr(j \text{ in } [a, c]) = \Pr(j \text{ in } [c, b]) = 0.5$$

And it follows that $\Pr(DS(j) = 0) = \Pr(DS(j) = 1) = 0.5$. *According to Definition 1,* $\forall i \in \mathcal{B}$ *, we have*

$$\eta(i) = \frac{|\Phi(i)|}{|\Theta(i)|} = \frac{\displaystyle\sum_{j=1}^{|\Theta(i)|-1} DS(j)}{|\Theta(i)|}, j \in \Phi(i) \tag{8}$$

By discrete random variable's operation, we can conclude that

$$\Pr(\eta(i) = \frac{1}{|\Theta(i)|}) = \Pr(\eta(i) = \frac{2}{|\Theta(i)|}) = \cdots$$

$$= \Pr(\eta(i) = \frac{|\Theta(i)|-1}{|\Theta(i)|}) = \frac{1}{|\Theta(i)|-1} \tag{9}$$

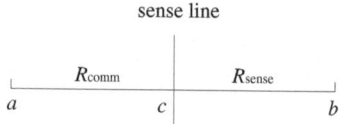

Fig. 4. Analysis of discovering neighbors ratio

Theorem 2: *Given sensor* i, $i \in \mathcal{B}$ *, let* H_0 *and* H_1 *be* $\eta(i)$ *'s thresholds, then the probability that sensor* i *is a BOUNDARY sensor is given by*

$$\Pr(BS(i) = BOUNDARY) = \sum_{H_0 < \eta(i) < H_1} \frac{1}{|\Theta(i)|-1}, \eta(i) \in Y \tag{10}$$

Proof: *Based on the Definition 2 and Lemma 1, the conclusion is obvious.*
Assuming group target's perimeter $L_{group\ target}$ *is long enough, we consider the BOUNDARY sensors constitute a curving band with the same length of outer and inner cures. Then a Max BOUNDARY Sensors Set's size can be computed by the following LEMMA.*

Lemma 2: *Given a Max BOUNDARY Sensors Set* $\mathcal{B}$ *, then its average size is given by*

$$\overline{|\mathcal{B}|} = \frac{L_{group\ target} \times W_{Boundary}}{\sqrt{1/\rho}} \tag{11}$$

where $L_{group\ target}$ *is group target's perimeter,* $W_{Boundary}$ *is boundary thickness, and* ρ *is the density of sensors.*

Proof: *Given a unit length of group target's boundary, the average number of sensors in this area can be calculated by $W_{Boundary}/d$, where d is the average distance between two sensors. Since $d = (1/\rho)^{1/2}$, then we can conclude that*

$$\overline{|\mathcal{B}|} = L_{\text{group target}} \times \frac{W_{\text{Boundary}}}{d} = \frac{L_{\text{group target}} \times W_{\text{Boundary}}}{\sqrt{1/\rho}} \tag{12}$$

Theorem 3: *Given sensor i, $i \in \mathcal{B}$, let H_0 and H_1 be $\eta(i)$'s thresholds, then the average size of Adjustable BOUNDARY Sensors Set B is given by*

$$\overline{|B|} = \frac{2L_{\text{group target}} \times \left\lfloor R_{\text{comm}}/\sqrt{1/\rho} \right\rfloor}{\sqrt{1/\rho}} \times \sum_{H_0 < \eta(i) < H_1} \frac{1}{\pi R_{\text{comm}}^2 \rho - 1}, \tag{13}$$

$$\overline{\eta(i)} \in Y' \quad \text{where } Y' = \{\frac{1}{|\Theta(i)|}, \frac{2}{|\Theta(i)|}, \cdots, \frac{|\Theta(i)|-1}{|\Theta(i)|}\}$$

Proof: *According to Theorem 1, Theorem 2 and Lemma 2, we can conclude that*

$$\overline{|B|} = \sum_{i=1}^{\overline{|\mathcal{B}|}} \overline{\Pr(BS(i) = BOUNDARY)}$$

$$= \overline{|\mathcal{B}|} \times \overline{\Pr(BS(i) = BOUNDARY)}|_{i \in \mathcal{B}} = \overline{|\mathcal{B}|} \times \sum_{H_0 < \eta(i) < H_1} \overline{\frac{1}{|\Theta(i)| - 1}}$$

$$= \frac{L_{\text{group target}} \times T_{\text{Boundary}}}{\sqrt{1/\rho}} \times \sum_{H_0 < \eta(i) < H_1} \frac{1}{\pi R_{\text{comm}}^2 \rho - 1} \tag{14}$$

$$= \frac{2L_{\text{group target}} \times \left\lfloor R_{\text{comm}}/\sqrt{1/\rho} \right\rfloor}{\sqrt{1/\rho}} \times \sum_{H_0 < \eta(i) < H_1} \frac{1}{\pi R_{\text{comm}}^2 \rho - 1}, \overline{\eta(i)} \in Y'$$

where $\overline{\Pr(BS(i) = BOUNDARY)}|_{i \in \mathcal{B}}$ is the average probability that sensor i, $i \in \mathcal{B}$ is a BOUNDARY sensor, and $\overline{|\Theta(i)|}$ is the average number of sensor i's neighbors.

4   Group Target Tracking Based on BSM

Based on our proposed BSM, a divide-merge group target tracking method is stated in this section.

In this method, the group tracking process is separated into two steps – boundary dividing and boundary merging. In the first step, the sensors that discover the boundary of a group target are divided into multiple clusters, and each cluster is responsible for tracking a partial boundary of the group target. In each cluster, there is a cluster head (CH) which gathers information from its cluster members (CM) and aggregates these data to form a local convex hull. Then, the aggregated data is sent back to the sink which is usually monitored by humans. In the second step, when sufficient information of local convex hulls is collected at the sink, it will execute the merging algorithm to combine those convex hulls into a global convex hull which is considered as the whole contour of the group target.

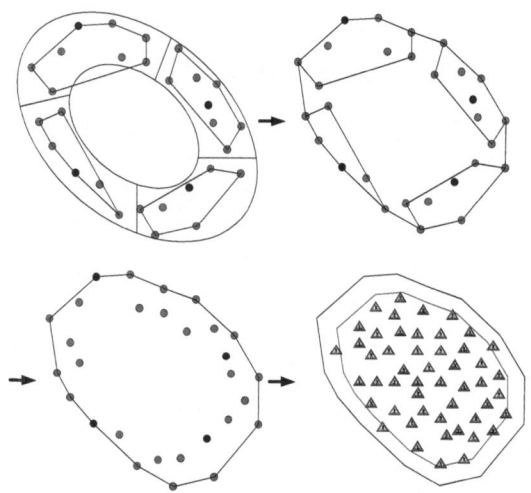

Fig. 5. Illustration of the divide-merge group target tracking method

When a group target is moving, some previously constructed clusters are destroyed and some new ones will be formed. In order to track the group target continuously, clusters must be maintained so that out-of-date clusters are eliminated and new clusters are dynamically created. Consider a newly formed cluster. As group target moving, some new sensors will join the cluster. Meanwhile, some old ones quit. So the topology of the cluster is changing and the original CH may not be able to continue playing as a cluster head. Here, we choose the sensor closest to the cluster's center as the new CH.

In Figure 5, the process of group target tracking based on BSM is illustrated. We note that several partial boundaries combine the whole boundary of group target.

5   Simulation and Verification

5.1   Simulation Setup

In the simulation, BSM sensors are randomly scattered with a uniform distribution in the monitoring region which is a rectangle area with the size of $200m \times 175m$. The communication radius and sense radius are changed according to the deployment density of sensors to guarantee enough coverage of the monitoring region.

A group target contained 400 individuals is simulated at the speed of $5m/s$. If there is any individual target gets into one sensor's circle whose radius is R_{sense}, the sensor will discover it without knowing the number of targets or their precise positions.

6 Impact Factors of Boundary Thickness and Tracking Sensors' Quantity

Based on the theoretical analysis in Section 3, we made a great deal of experiments on the factors that impact on boundary thickness.

Figure 6 illustrates the impaction of H_0 and H_1. With the difference between H_0 and H_1 getting smaller, the boundary thickness is getting thinner accordingly. This is reasonable according to Theorem 3.

The impaction of communication radius R_{comm} on boundary thickness is given in Figure 7. We can find the boundary thickness broadens with R_{comm} increasing. This is because a sensor has more neighbors, and then it has bigger chance to become a BOUNDARY sensor.   Theorem 1 predicts the trend.

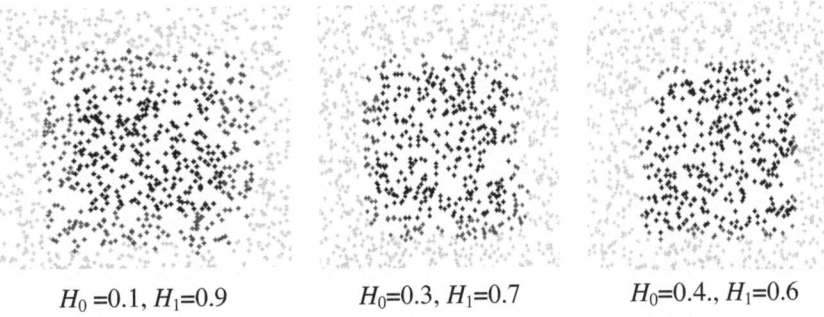

$H_0 =0.1, H_1=0.9$ $H_0=0.3, H_1=0.7$ $H_0=0.4., H_1=0.6$

Fig. 6. Boundary thickness changes by vary H0 and H1

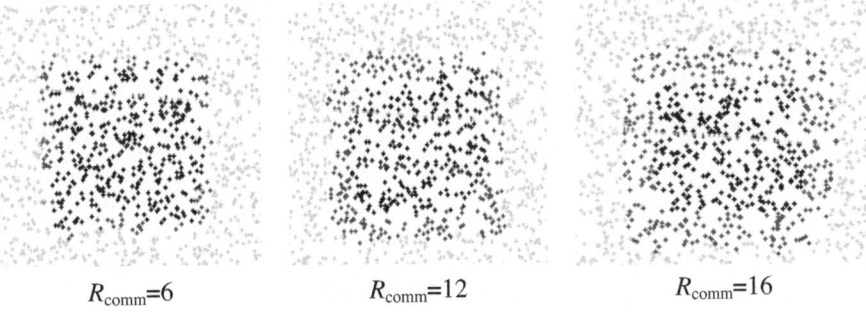

$R_{comm}=6$ $R_{comm}=12$ $R_{comm}=16$

Fig. 7. Boundary thickness changes by vary Rcomm

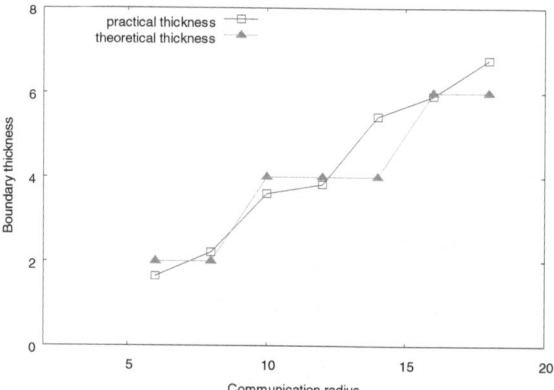

Fig. 8. Comparison between practical and theoretical boundary thickness

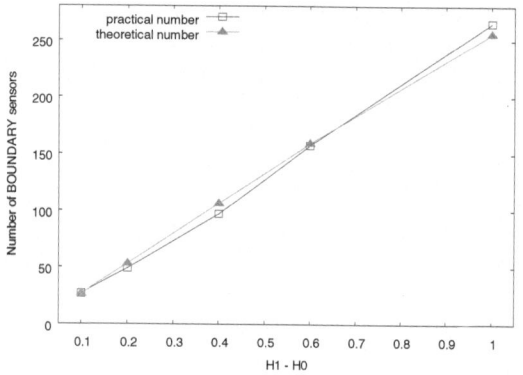

Fig. 9. Comparison between practical and theoretical BOUNDARY sensors' quantity

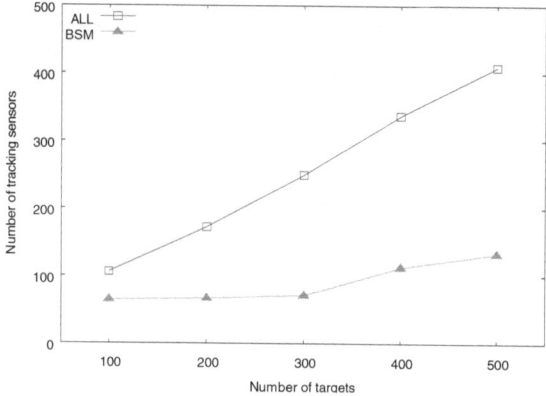

Fig. 10. Comparison between BSM based tracking sensors' number and the one that all discovering sensors are involved in tracking

In Figure 8 and Figure 9, the comparisons between practical and theoretical results are illustrated, from which it is obvious that our theoretical results are closed to the practical situation.

In Figure 10, the tracking sensors' quantities comparison between BSM based tracking (BSM tracking) and tracking involved all discovering sensors (ALL tracking) is illustrated. With the expanding of group target's scale, the tracking sensors number increases faster in ALL tracking; While in BSM tracking, the number does not increase significantly. The reason is that the number of tracking sensors varies directly with the group target's area in ALL tracking, and varies directly with the group target's perimeter in BSM tracking.

7   Tracking Performance

In Figure 11, keeping the difference between H_0 and H_1 fixed, we can find the boundary become tighter with H_0 and H_1 increasing. The reason is that when H_0 and H_1 increase, some INNER sensors closed to the boundary turn into BOUNDARY sensors, while some outer BOUNDARY sensors turn into OUTER sensors. As a result, it's helpful to make H_0 and H_1 a little higher. However, if the H_0 and H_1 turn too high, the localization result may be smaller than the real contour.

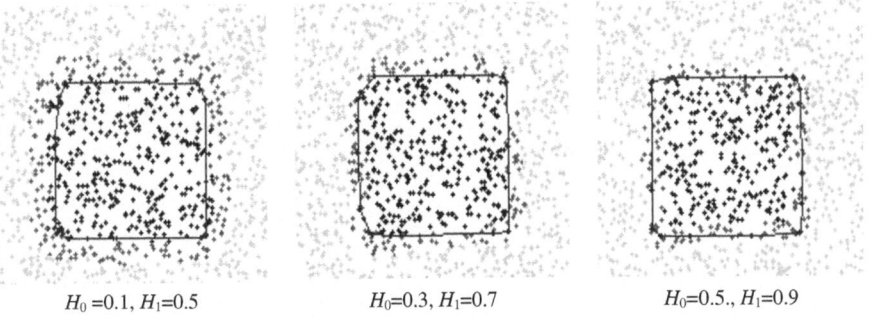

$H_0 =0.1, H_1=0.5$ $H_0=0.3, H_1=0.7$ $H_0=0.5., H_1=0.9$

Fig. 11. H0 and H1 s impactions on localization precision

8   Conclusion

In this paper, we suggested a boundary sensing model - BSM used for group target tracking. We mapped the boundary sensing problem to the test of whether a sensor's discovering neighbors ratio is within boundary thickness thresholds. And we derived analytical expressions for the probability that a sensor turns into a BOUNDARY sensor. After the sensor network is deployed uniformly, we showed that the number of tracking sensors in BSM only depends on our proposed thresholds. Through simulation, we verified our theoretical results and confirmed our proposed group target tracking method based on BSM eliminates the tracking sensors' quantity significantly under the premise of precision.

References

1. Zhu, X., Sarkar, R., Gao, J., Mitchell, J.S.B.: Light-weight Contour Tracking in Wireless Sensor Networks. In: Proceedings of IEEE INFOCOM 2008, pp. 1849–1857 (2008)
2. Aslam, J., Butler, Z., Constantin, F., Crespi, V., Cybenko, G., Rus, D.: Tracking a Moving Object with a Binary Sensor Network. In: ACM Sensys 2003, Los Angeles, California, USA, pp. 150–161 (2003)
3. Shrivastava, N., Mudumbai, R., Madhow, U., Suri, S.: Target tracking with binary proximity sensors: Fundamental limits, minimal descriptions, and algorithms. In: Proceedings of ACM SenSys (2006)
4. Chen, W., Hou, J.C., Sha, L.: Dynamic clustering for acoustic target tracking in wireless sensor networks. IEEE Transactions on Mobile Computing 3(3), 258–271 (2004)
5. Ji, X., Zha, H., Metzner, J.J., Kesidis, G.: Dynamic Cluster Structure for Object Detection and Tracking in Wireless Ad-Hoc Sensor Networks. In: IEEE International Conference on Communications, ICC 2004, Paris, France, June 20-24 (2004)
6. Cao, D., Jin, B., Cao, J.: On Group Target Tracking with Binary Sensor Networks. In: Proceedings of the 5th IEEE International Conference on Mobile Ad Hoc and Sensor System, MASS, pp. 334–339 (2008)
7. Gui, C., Mohapatra, P.: Power Conservation and Quality of Surveillance in Target Tracking Sensor Networks. In: Proc. of the 10th Annual International Conference on Mobile Computing and Networking, pp. 129–143 (September 2004)
8. Jiang, C., Dong, G., Wang, B.: Detecting and Tracking of Region-based Evolving Targets in Sensor Networks. In: IEEE ICCCN 2005, San Diego, California (2005)

Green Femtocell Networking with IEEE 802.16m Low Duty Operation Mode

Ching-Chun Kuan[1], Guan-Yu Lin[2], and Hung-Yu Wei[1,2,*]

[1] Department of Electrical Engineering, National Taiwan University
[2] Graduate Institute of Communication Engineering, National Taiwan University
hywei@cc.ee.ntu.edu.tw

Abstract. Femtocell technique has become a radical approach to improve network capacity and coverage. To efficiently reduce interference between femto BS and other BSs, low duty operation mode (LDM) is deemed as a good method to mitigate the interference. In this work, we develop a mechanism that incorporates the power saving mechanism in both femto BS (LDM mode) and MSs (sleep mode). To correspond to the current IEEE 802.16m system architecture, we build a practical system model and analyze the performance of the proposed modification. The simulation shows that with simple sleep-cycle operation, our mechanism can achieve both a high sleep ratio and an acceptable low delay without generating more control overhead than original system requirement. With the above characteristics, our mechanism provides a solid basis to create a green communication system.

Keywords: femtocell, low-duty operation mode, sleep mode.

1 Introduction

Along with the raising environmental awareness, green communications capture the public's attention. X. Wang *et al.* [1] provided a comprehensive survey of green techniques for mobile networks. The authors pointed out the importance of femtocell in green communication. The key for communications to go green is to redesign the network's infrastructure into a more efficiency one. Among the existing technologies, power-saving design is especially crucial in wireless communication that needs to be addressed in this issue. IEEE 802.16m [2] specifies two power saving mechanisms : sleep mode for mobile station (MS) and low duty operation mode (LDM) for femto BS. The former one, sleep mode in 802.16m, inherits the design in 802.16e, and has further improvements to the power efficiency. The latter one, LDM on femtocell [3][4], supports the function of scheduling availability intervals, i.e. puts the femto BS to a passive mode that activated on air interface when receiving a connection request. This reduces the system resources that are allocated for the use of femto BS. For a light-loaded femto BS, LDM operation enables power saving and interference mitigation with

---

* Corresponding author.

Joel J.P.C. Rodrigues et al.: (Eds.): GreeNets 2011, LNICST 51, pp. 37–50, 2012.
© Institute for Computer Sciences, Social Informatics and Telecommunications Engineering 2012

tolerable delay performance. In short, LDM shows great promises to efficiently improve the network capacity with its strength of low cost, small coverage and low transmission power. However, in the current standard, an integrated design of both LDM and sleep mode is not yet supplied, which undermines the power-saving performance of LDM when the femto BS serving multiple MSs at the same time. This is because the MSs have diverse traffic patterns that place different timings of listening window. The objective of this paper is to achieve an efficient power use for femto BS and MSs by means of integrating LDM and sleep mode.

2   Related Work

Much research has conducted to evaluate the performance of PSC (power saving class) Type I and Type II in IEEE 802.16e. The first theoretical model of sleep mode is proposed by Xiao [5]. Considering downlink traffic, this work modeled PSC Type I with Markov Chain, in which states represents the sizes of sleep cycle. Zhang and Fujise [6] proposed an analytical model considering both downlink and uplink traffic, which also addresses the effect of traffic direction ratio : the heavier uplink traffic, higher the power consumption. Han and Choi [7] developed a model considering both sleep and awake mode, providing detailed discussions about the effect of parameters. Kong et al. [8,9,10] analysed power consumption and traffic delay in PSC Type I and Type II, proposing a Markov Decision Process (MDP) framework to select the best PSC for MS.

Some research offers performance analysis on the enhancement of sleep mode for IEEE 802.16m. Baek et al. provided a 2-dimensional Markov chain to model 802.16m with listening window extension in [11]. In [12], Jin et al. proposed a mechanism considering both realtime and non-realtime traffic. Recently, some research conducted on innovative designs for 802.16m architecture. In [13], Kalle et al. the authors addressed the issue of increasing number of PSC connections worsening the power saving efficiency. They proposed a mechanism that manages joint power class to aggregate individual traffic. This work improves the power efficiency on mobile phones, whereas ours improves the power saving efficiency both mobile phones and femto BS at the same time.

The mentioned works primarily address power consumption issue on MS, while similar research on femto BS is not yet supplied. The objective of this paper is to enhance power efficiency on both femto BS and MS. To achieve that goal, we propose a novel sleep mode scheme that can enhances the power efficiency of femto BS significantly, providing a complete performance analysis on both femto BS and MS.

3   Power Saving Mechanism

3.1   Sleep Mode in 802.16m

The main reason of power-saving is to prolong the lifetime of mobile phone. Furthermore, the energy-consumption in MSs rises rapidly. Hence, power-saving mechanism of MS is essential in green communication. In the IEEE 802.16m standard, MSs can support Sleep Mode for saving power. Sleep Mode is a state in which an MS conducts periods of absence from the serving BS air interface. During the activation of Sleep Mode, the MS is provided with series of Sleep Cycles that consist of a Listening Window followed by a Sleep Window. During Listening Window, the MS is expected to receive all downlink transmissions same way as in Active Mode. During Sleep Window, the BS shall not transmit downlink traffic to the MS, therefore the MS may power down some physical operation components.

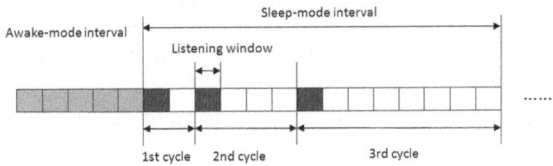

Fig. 1. An example of the 802.16m PSC Type I operation

Power Saving Class of Type I

PSC type I is recommended for connections of Best effort and Non-Real Time service type, as illustrated in Fig.1. The sleep cycle begins with a listening window, and then a sleep window follows. The length of Listening Window T_L is fixed. During the listening window, MS should wake and listen to the channel. At first, the sleep cycle is the Initial (Minimal) Sleep Cycle T_{min}. The sleep cycle will grows doubly until it reaches the Final (Maximum) Sleep Cycle T_{max} if the traffic indication is negative. To sum up, the MS and BS shall update the length of the Sleep Cycle when traffic indication is negative as follows:

$$\text{Current Sleep Cycle} = \min\,(2 \times \text{Previous Sleep Cycle},$$
$$\text{Final Sleep Cycle})$$

On the other hand, if the traffic indication is positive, then MS and BS can negotiate the next sleep cycle length. The general case is that MS reset the sleep cycle to Initial Sleep Cycle when the traffic indicator is positive.

Power Saving Class of Type II

When the Final Sleep Cycle is equal to the Initial Sleep Cycle, the length of sleep cycle is fixed. Those kind of Power Saving Class are PSC type II. PSC type II recommended for connections of UGS and RT-VR service connection. Due to the fixed length of sleep cycle, the packet delay is restricted in a range. Compare to Type I, MS pays more power for reducing delay in Type II.

3.2   Low Duty Mode in Femtocell BS

Besides the normal operation mode, femto BSs may support low-duty operation mode, in order to reduce interference to neighbor cells and save power. In LDM mode, femto BS is (a)periodically active and inactive on air interface. An on-off cycle, named low duty cycle, consists of available intervals (AIs) and unavailable intervals (UAIs). When in UAIs, no transmission is performed on air interface, whereas in AIs femto BS can be active on air interface, such as doing ranging, signaling procedures, or data traffic transmission. A LDM pattern is composed of a sequence of AIs and UAIs.

For illustrating the relationship between MS and BS, we shows the following equation:

$$A(i)_{BS} = \begin{cases} 1 & i \text{ is a multiple of } T_{LDM}, \\ \cup_{j=1}^{N} A(i)_{MS_j} & \text{otherwise} \end{cases}$$

where T_{LDM} is the default LDM cycle of femto BS. $A(i)_{BS}$ represents the mode of BS in the i-th frame. If BS is awake in the i-th frame, then $A(i)_{BS} = 1$. Otherwise, $A(i)_{BS} = 0$. Similarly, $A(i)_{MS_j}$ represent the mode of the j-th MS in the i-th frame.

According to [14], a femto BS can enter LDM mode under two conditions : either all MSs served by this femto BS are in sleep or idle mode, or no MS is in the service range of the femto BS. That is, only when all MSs are light-loaded can femto BS be in LDM mode. Besides, [2] specifies that femto BS can freely schedule AI periods based on operational requirement. In other words, current documentations provide the flexibility of design LDM pattern design.

4   Proposed Sleep Mode Scheme

As mentioned previously, the benefits of LDM are power-saving and interference-mitigation. And both of these merits are based on fewer available intervals. However, when the number of MSs attached to the the same femto BS increases, the power-saving efficiency of femto BS degrades. The fact is illustrated in Fig.2, in which 3 MSs in the Sleep Mode are attached to the same femto BS. Each MS has the parameters $T_{min} = 2$ frames and $T_{max} = 8$ frames. And the default LDM cycle of femto BS are 8 frames. We assume the Sleep Cycle will be reset to Initial Sleep Cycle if the traffic indicator is positive. In Fig.3, we can observe that the length of Sleep Cycle in each MS begins different as the packets arrives, which means the Listening Windows of each MSs are in different frames. Femto BS still wakes frequently even though there is no traffic after the third LDM cycle. Therefore, the power-efficiency of BS becomes lower. An intuitive solution is to synchronize the Listening Window of each MS by choosing the same Sleep Cycle. Because the amount of data traffic is small in Sleep Mode, so femto BS is supposed to serve all the MSs at one frame. However, the delay tolerance of each MS may be different. It is not feasible to set the same Sleep Cycle of each MS. To enhance the power-efficiency of BS and take account of different traffic type, we propose a new sleep mode which is based on the IEEE 802.16m standard. The proposed solution is to minimize necessary AIs by synchronizing Listening Window of MSs at the same frame time. Hence, the Available Intervals of femto BS can be reduced.

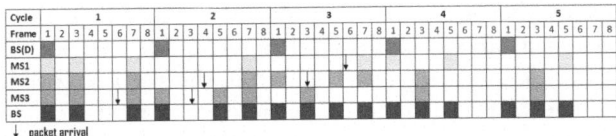

Fig. 2. An example of original PSC Type I operation with downlink traffic

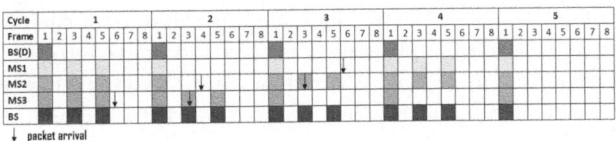

Fig. 3. An example of our sleep mode operation with downlink traffic

As we mentioned before, when the number of MSs attached to the the same femto BS increase, the power saving efficiency of femto BS decreases. The fact is illustrated in Fig.2, in which 3 MSs in sleep mode (PSC Type I) are attached to the same femto BS. Each MS has the parameters $T_{min} = 2$ frames and $T_{max} = 8$ frames. And the default LDM cycle of femto BS are 8 frames. If the traffic indicator is positive, then reset the sleep cycle to initial sleep cycle. Observe that the length of sleep cycle in each MS comes to be different as the packets arrives, which means listening windows of MSs are in different frame time. Therefore, femto BS has to apply more AIs and thus leading to lower power-saving efficiency in BS. The proposed solution is to minimize necessary AIs by aligning listening window of MSs on the same frame time. Here three parameters are listed below :

- T_{MS_i}: current sleep cycle size of the i-th MS.
- T_{total}: sum of historical and current sleep cycle size.
- T_{BS}: the interval from current AI to next AI of femto BS.

Algorithm 1. proposed sleep cycle scheme

---

if traffic indicator is negative **then**
   if T_{total} mod $2T_{MS} == 0$ **then**
      $T_{MS} \leftarrow min(2T_{MS}, T_{max})$
   else
      $T_{MS} \Leftarrow T_{MS}$
   end if
else {traffic indicator is positive}
   $T_{MS} \Leftarrow T_{min}$
end if

Algorithm 1 is our sleep mode scheme algorithm. The algorithm should be executed in listening window for each MS whenever a new sleep cycle starts.

This algorithm will determine the next sleep cycle length. If the traffic indicator is negative, then MS calculates T_{total} modulo T_{MS}. If the value is 0, then $T_{MS} = min(2T_{MS}, T_{max})$. Otherwise, maintain the sleep cycle until next listening window. If the traffic indicator is positive, then reset T_{MS} to initial sleep cycle.

We modify the original design when the traffic indicator is negative. In the original IEEE 806.16m standard, the sleep cycle will be double when traffic indicator is negative. In our scheme, we add a restriction for increasing sleep cycle. In this way, our algorithm can ensure the listening windows of each MS will locate in the $(T_{MS} \times K + 1)$-th frame. Where K is zero or positive integer. Therefore, we can ensure that $T_{BS} = min\{T_{MS_i}\}$. However, in the original sleep mode scheme, because each MS has different traffic pattern, their listening windows locate disorderly. Hence, T_{BS} is less than or equal to $min\{T_{MS_i}\}$. It is obvious that our sleep mode scheme outperforms than the original sleep mode in the perspective of femto BS due to longer T_{BS}.

Fig.3 illustrates our sleep mode scheme with the same traffic pattern in Fig.2. In Fig.2, the total available intervals are 16 frames; however, in Fig.3, the total AIs are only 13 frames by listening window alignment. Please note that compared with PSC Type I, more power of femto BS but less power of MS does proposed alignment procedure save. This is because proposed scheme put one more restriction on MS's sleep cycle doubling.

5   Analytical Model

5.1   Assumptions

For simplicity, we assumes the packet of each MS arriving at femto BS follows Poisson process with arrival rate λ. Besides, the transmission rate is 1 packet/frame in listening window.

5.2   Markov Chain Model

We propose a two-dimensional Markov chain analysing the performance of MS. Now we define the state notation $S_{x,y}$. The variable x represents the sleep cycle in the state. For example, Sleep cycle T_{min} corresponds to $x = 1$; $2 \times T_{min}$ corresponds to $x = 2$; etc. We denote T_i as the sleep cycle in the states whose x variable is i. Hence, $T_i = 2^{i-1} \times T_{min}$. There are $\frac{T_{max}}{T_i}$ states whose x variable is i. The parameter y correspond to the remainder of T_{total} divided by T_{MS}. In state $S_{x,y}$, the sleep cycle is T_x, and the remainder of T_{total} divided by T_{MS} is $T_x \times (y - 1)$. We denote the steady state probability in $S_{x,y}$ as $\pi_{x,y}$. For convenience, we use the following notation:

- N_{NS} represents the total number of sleep cycle size.
- N_X represents the number of states whose x variable is X.

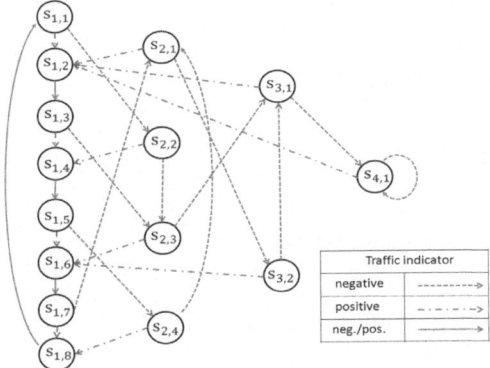

Fig. 4. A state diagram of MS with $T_{min} = 1$ and $T_{max} = 8$

According to the above definition, we can get $N_{NS} = log_2(\frac{T_{max}}{T_{min}}) + 1$ and $N_X = \frac{T_{max}}{T_X}$.

Fig.4 shows the state diagram of an MS with $T_{min} = 1$ and $T_{max} = 8$. We can see that the sleep cycle will be reset to initial sleep cycle if the traffic indicator is positive. On the other hand, whether the sleep cycle will maintain or increase depends on the state when the traffic indicator is negative.

After constructing the Markov Chain model, we derive the state transition probability. We define transition probability $P_{m,n,r,s} = P(S_{m,n}$ in i+1-th cycle $|S_{r,s}$ in i-th cycle). And we classify the different cases by the parameters r and s as follow:

$r = 1$, s is odd number,

$$\begin{cases} P_{r,s+1,r,s} = 1 - e^{-\lambda T_r} \\ P_{r+1,mod(\frac{s+1}{2},N_{r+1})+1,r,s} = e^{-\lambda T_r} \end{cases} \tag{1}$$

$r = 1$, s is even number,

$$P_{r,s+1,r,s} = 1 \tag{2}$$

$N_{NS} > r > 1$, s is odd number,

$$\begin{cases} P_{r+1,mod(\frac{s+1}{2},N_{r+1})+1,r,s} = e^{-\lambda T r} \\ P_{1,(s-1)\times\frac{T_r}{T_1}+2,r,s} = 1 - e^{\lambda T r} \end{cases} \tag{3}$$

$N_{NS} > r > 1$, s is even number,

$$\begin{cases} P_{r,mod(s+1,N_r),r,s} = e^{-\lambda T r} \\ P_{1,(s-1)\times\frac{T_r}{T_1}+2,r,s} = 1 - e^{-\lambda T r} \end{cases} \tag{4}$$

$r = N_{NS}$,

$$\begin{cases} P_{r,s,r,s} = 1 - e^{-\lambda T_r} \\ P_{1,2,r,s} = e^{-\lambda T_r} \end{cases} \tag{5}$$

According to equation (1)-(5), we can find the balance equation of the Markov chain as follows:

$r = 1$, s is odd number,

$$\pi_{r,s} = \pi_{r,s-1} \tag{6}$$

$r = 1$, s is even number,

$$\pi_{r,s} = \pi_{r,s-1} \times \left(1 - e^{-\lambda T_1}\right) + \sum_{m,n}^{s-2=\frac{T_m}{T_1} \times (n-1)} \pi_{m,n} \times \left(1 - e^{-\lambda T_m}\right) \tag{7}$$

$N_{NS} > r > 1$, s is odd number,

$$\pi_{r,s} = \pi_{r-1,mod(2s-3,T_{r-1})} \times e^{-\lambda T_{r-1}} + \pi_{r,s-1} \times e^{-\lambda T_r} \tag{8}$$

$N_{NS} > r > 1$, s is even number,

$$\pi_{r,s} = \pi_{r-1,mod(2s-3,T_{r-1})} \times e^{-\lambda T_{r-1}} \tag{9}$$

$r = N_{NS}$,

$$\pi_{r,1} = \pi_{r-1,1} \times e^{-\lambda T_{r-1}} \tag{10}$$

With the normalization condition

$$\sum_{m=1}^{N_{NS}} \sum_{s=1}^{N_r} \pi_{r,s} = 1 \tag{11}$$

We can solve the steady state probabilities by equation (6)-(11).

6   Performance Evaluation

To evaluate the performance of sleep mode, we define the steady state probability $\pi_i = \sum_{j=1}^{N_i} \pi_{ij}$.

6.1   Sleep Ratio of MS

Definition 1. We define *sleep ratio* of MS ϕ as the sum of sleep window interval divided by the sum of sleep cycle length during total sleep mode interval.

Please note that even when an MS operates in sleep mode, it should wake and listen channel aperiodically. Hence the *sleep ratio* of MS represents the ratio of real power saving duration in sleep mode. Our goal is trying to enhance the ratio without degrading other performance metrics.

we can evaluate the sleep ratio as follows:

$$\phi = \frac{\sum_{i=1}^{N_S} \pi_i (T_i - T_L)}{\sum_{i=1}^{N_S} \pi_i T_i} \tag{12}$$

The denominator represents the normalized length of sleep cycle, and the numerator represents the normalized length of sleep window. By equation (12), we can determine the sleep ratio with the steady state probability.

6.2   Sleep Ratio of Femto BS

Definition 2. We define *sleep ratio* of BS θ as the sum of active intervals divided by the total LDM intervals.

The *sleep ratio* of BS represents the ratio of power saving duration of a femto BS during Low-duty operation mode. Similar to the *sleep ratio* of MS, we hope to increase this ratio as much as possible.

Definition 3. We define 1 *unit* $= T_{min}$ frames.

For evaluating sleep ratio of BS, we use following notation:

- $begin(i)$: it is a set that the sleep cycle begins in the i-th unit
- $begin_L(i, s)$: it is a subset of $begin(i)$. The sleep cycle of elements should not exceed s units.
- $B(i)$: total steady state prob. in $begin(i)$
- $B_L(i, s)$: total steady state prob. in $begin_L(i, s)$
- $P_{MS}(i)$: the prob. that a sleep cycle begins at the i-th unit in MS; in other word, it is the prob. that listening window occurs at the i-th unit in BS.
- $P_{BS}(i)$: the prob. that a sleep cycle begins at the i-th unit in BS; in other word, it is the prob. that listening window occurs at the i-th unit in BS.

Our goal is finding the sleep ratio of BS. But it is hard to get the value directly, so we find $P_{MS}(i)$ first, and then we can calculate $P_{BS}(i)$ and sleep ratio of BS easily. The prob. that cycle begins at the i-th unit is equal to the prob. that cycle ends at the i-1 th unit. If no cycle end is the k-th unit, then we can assure that there is a long sleep cycle begins at the k-s th unit and the cycle length exceed s units. According to the above knowledge, we derive the general form of $P_{MS}(i)$ as follows:

$$P_{MS}(i) = P_{MS}(k)\frac{B_L(k, i - k)}{B(k)} \tag{13}$$

where k is the largest integer which satisfied the following conditions:

$$k \bmod 2^n = 0, n \in N; k < i$$

k is the key position which dominates the behavior of the i-th unit. As long as the sleep cycle doesn't exceed i-k units at the k-th unit, some cycle is sure to begin at the i-th unit. We only need to consider $\frac{T_{max}}{T_{min}}$ units (or T_{max} frames), because T_{max} is the largest sleep cycle. Please note that $P_{MS}(1) = 1$.

We assume a femto BS serves N identical MSs. Hence we can get $P_{BS}(i) = 1 - (1 - P_{MS}(1))^N$. Finally, we can calculate sleep ratio of BS by the following equation:

$$\theta = 1 - \frac{\sum_{i=1}^{T_{max}} P_{BS}(i)}{T_{max}} \tag{14}$$

Where $\sum_{i=1}^{T_{max}} P_{BS}(i)$ means the average active intervals in T_{max} frames.

6.3   Packet Delay

Definition 4. We define *packet delay* D is the interval between packet arriving and successfully received.

For simplicity, we assume packet arrivals follows the Poisson distribution. And there is no transmission error. Thus, M/G/1 queue with vacation is the best model to describe the behavior of traffic. According to the Pollaczek-Khinchin formula for M/G/1 with vacations :

$$\begin{cases} W = \frac{\lambda \bar{X}^2}{2(1-\rho)} + \frac{\bar{V}^2}{2\bar{V}} \\ T = \frac{1}{\mu} + W \end{cases} \tag{15}$$

where the notations are defined as Table 1.

Table 1. Notations of M/G/1 model

Notation	meaning
λ	packet arrival rate
W	packet waiting time in queues
T	total waiting time
X	service time, i.e. the transmission time
μ	service rate, i.e. the packet transmission rate
ρ	the ratio of λ to μ
V	vacation interval, i.e. the sleep window size of MS
$\bar{V}$	the first moments of the vacation interval
$\bar{V}^2$	the second moments of the vacation interval

According to [10], a compensation term d_s should be added to total delay. Where $d_s = \frac{V}{2}\rho$. So the mean packet delay $E[D]$ can be evaluated by

$$E[D] = T + d_s \tag{16}$$

Obviously, the transmission time X is $\frac{1}{\mu}$, our work is to find $\bar{V}$ and $\bar{V}^2$. Again we can find those by steady state probability as follows:

$$\begin{cases} \bar{V} = \sum_{i=1}^{N} \pi_i V_i \\ \bar{V}^2 = \sum_{i=1}^{N} \pi_i V_i^2 \end{cases} \tag{17}$$

where V_i is the length of Sleep Window in the states whose x variable is i, hence $V_i = T_i - T_L$.

7   Simulation Results and Discussions

To verify our theoretical model, we write an event-driven simulator which implemented in Matlab code. At first, we compare the simulation results of MS in the proposed sleep mode scheme to the theoretical value.

Table 2. Parameter setting

Parameters	Setting
Initial Sleep Cycle	2 Frames
Final Sleep Cycle	512 Frames
Listening Window	1 Frame
Transmission Rate	1 per frame

The simulator parameter setting of MS is shown in Table 2 and the default LDM cycle of femto BS are 512 frames.

Fig.5 has two parts. The left part shows the sleep ratio of MS based on proposed sleep mode in different arrival rate. We can see that if the packet arrival rate increases, the sleep ratio decreases. The right part shows the mean packet delay of MS based on the novel sleep mode in different arrival rate. The mean packet delay decreases when the arrival rate increases. The reason is that the sleep cycle will be reset to initial sleep cycle when traffic arrived. Fig.6 shows the sleep ratio of BS and MS in different number of MS. The sleep ratio of BS reduces when number of MSs increase. It is clear that the analysis and the simulation match each other very well.

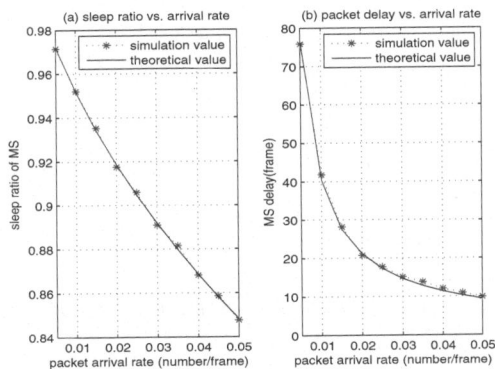

Fig. 5. The simulation result and theoretical value in our sleep mode scheme

To reveal the advantage of our design, we choose the original PSC type I sleep mode as the control group. The performance metrics we considered are the sleep ratio of BS, the sleep ratio of MS and the mean packet delay. We define the sleep ratio of BS as the ratio of total unavailable interval to total low-duty mode interval. In Fig.7 and Fig.8, we assume a femto BS serving 5 MSs. In proposed scheme and original scheme, each MS has the same parameters as Table 2. Fig.7 shows sleep ratio of MS and femto BS according to packet arrival rate. We can see that the sleep ratio of BS in proposed scheme is higher than that of original scheme. Please recall that proposed algorithm can align the listening window of each MS. Hence, the Active Intervals of femto BS can be reduced. On the other hand, the sleep ratio of MS in proposed scheme is lower than that of original

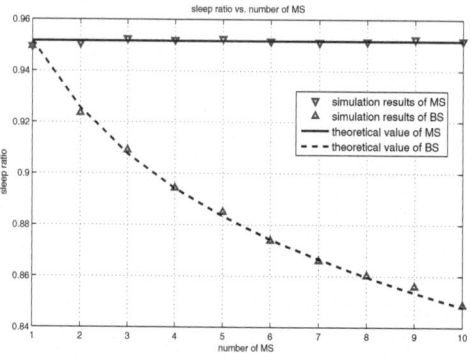

Fig. 6. Sleep ratio of MS and femto BS according to number of MS

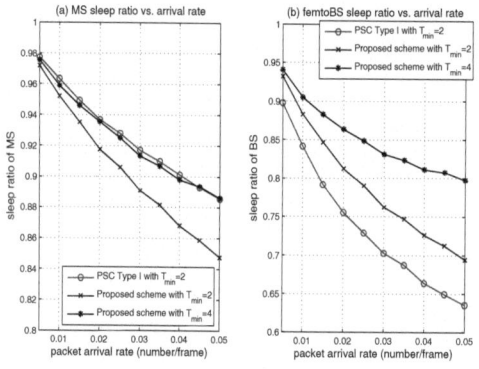

Fig. 7. Sleep ratio of MS and femto BS according to packet arrival rate

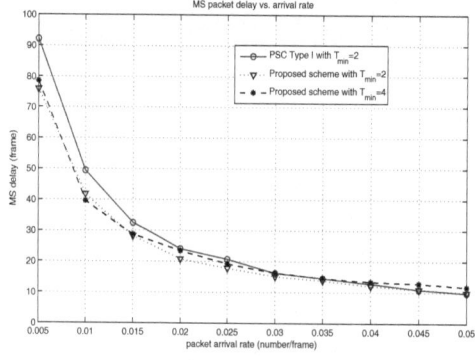

Fig. 8. Mean packet delay according to packet arrival rate

scheme. The reason is that we put a restriction of increasing sleep cycle, so the sleep cycle in our scheme is less than or equal to the original scheme. Fig.8 shows the mean packet delay with different arrival rate. In proposed scheme with $T_{min} = 2$, the packet delay is always lower than that of original scheme because its sleep cycle is less than or equal to that of original scheme.

Now we have a brief summary : in the same parameters, our scheme has higher power sleep ratio of BS and has lower mean packet delay but cost more power in MS. Please note that there is a trade-off relationship between power and delay. And we can save more power in exchange of longer delay by adjusting the parameters. For example, if we want to increase the sleep ratio of MS, then we can choose a bigger initial sleep cycle. Due to the larger initial sleep cycle, both the sleep ratio in MS and mean packet delay are increased. According to Fig.7 and Fig.8, the curves of MS sleep ratio and MS packet delay between "PSC Type I with $T_{min} = 2$" and "Proposed scheme with $T_{min} = 4$" are very close. However, the sleep ratio of BS in proposed scheme is much higher than that of original scheme. It is obviously that our sleep mode scheme outperforms than original scheme.

We illustrate the relationship between sleep ratio of BS and number of MS in Fig.9. It is clear that if femto BS serves more MSs, the sleep ratio of BS in low duty mode will decrease. The downtrend of sleep ratio in proposed scheme is much gentler than the original scheme. Because our scheme can efficiently align the listening window of each MS. To sum up, our scheme can save a great deal of power in femto BS and still maintain the performance of MS. And the advantage of our scheme become more significant when the number of MS increases.

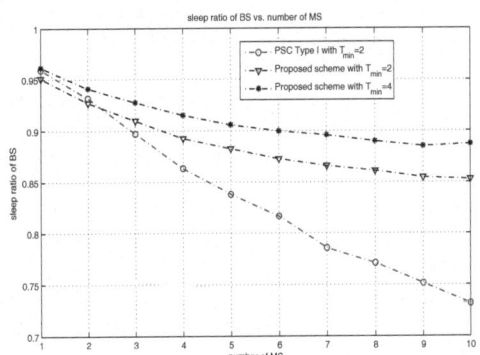

Fig. 9. Sleep ratio of BS according to number of MS. (Packet arrival rate = 0.01 per frame)

8   Conclusions

In this paper, we develop a power saving scheme to satisfy the criteria of green communications. As addressed, power efficiency can be further improved by means of incorporating two kinds of power-saving mechanisms, i.e. LDM for femto BS and sleep mode for MS. We provide a complete performance analysis of sleep ratio and packet delay by applying Markov Chain model and M/G/1 queueing model. Simulation results show that the proposed scheme achieves higher power efficiency in femto BS with a negligible delay increase. Our main contribution is to propose a simple and implementable mechanism to achieve both power efficiency and interference mitigation in femtocell network. Such mechanism takes only some simple distributed calculation with a slight overhead increase to achieve power efficiency. To the authors best knowledge, this is the first work incorporating both LDM operation and sleep mode into the power efficiency design.

Acknowledgements. The authors are in part supported by National Science Council, National Taiwan University and Intel Corporation under Grants NSC100-2219-E-002-016-,   NSC100-2219-E-009-027-,   NSC99-2911-I-002-001, 99R70600,and 10R70500.

References

1. Wang, X., Vasilakos, A., Chen, M., Liu, Y., Kwon, T.: A survey of green mobile networks: Opportunities and challenges. In: Mobile Networks and Applications, pp. 1–17 (2011)
2. IEEE Std 802.16m-2011, IEEE Standard for Local and metropolitan area networks, Part 16: Air Interface for Broadband Wireless Access Systems: Advanced Air Interface (2011)
3. Chandrasekhar, V., Andrews, J., Gatherer, A.: Femtocell networks: a survey. IEEE Communications Magazine 46(9), 59–67 (2008)
4. Claussen, H., Ho, L.T.W., Samuel, L.G.: An overview of the femtocell concept. Bell Labs Technical Journal 13(1), 221–245 (2008)
5. Xiao, Y.: Energy saving mechanism in the IEEE 802.16 e wireless MAN. IEEE Communications Letters 9(7), 595 (2005)
6. Zhang, Y., Fujise, M.: Energy management in the IEEE 802.16 e MAC. IEEE Communications Letters 10(4), 311 (2006)
7. Han, K.: Performance analysis of sleep mode operation in IEEE 802.16 e mobile broadband wireless access systems
8. Kong, L., Tsang, D.H.K.: Performance study of power saving classes of type I and II in IEEE 802.16 e. In: Proceedings 2006 31st IEEE Conference on Local Computer Networks, pp. 20–27 (2006)
9. Kong, L., Tsang, D.H.K.: Optimal selection of power saving classes in IEEE 802.16 e. In: IEEE Wireless Communications and Networking Conference, WCNC 2007, pp. 1836–1841 (2007)
10. Kong, L., Wong, G.K.W., Tsang, D.H.K.: Performance study and system optimization on sleep mode operation in IEEE 802.16 e. IEEE Transactions on Wireless Communications 8(9), 4518–4528 (2009)
11. Baek, S., Son, J.J., Choi, B.D.: Performance Analysis of Sleep Mode Operation for IEEE 802.16 m Advanced WMAN
12. Jin, S., Choi, M., Choi, S.: Performance Analysis of IEEE 802.16 m Sleep Mode for Heterogeneous Traffic. IEEE Communications Letters 14(5), 1 (2010)
13. Kalle, R.K., Raj, M., Das, D.: A novel architecture for IEEE 802.16 m subscriber station for joint power saving class management. In: COMSNETS (2009)
14. Low-duty Operation Mode. IEEE 802.16m System Description Document (SDD), p. 148 (2010)

An Android Multimedia Framework
Based on Gstreamer

Hai Wang[1], Fei Hao[2], Chunsheng Zhu[3],
Joel J.P.C. Rodrigues[4], and Laurence T. Yang[3]

[1] School of Computer Science, Wuhan University, Wuhan, China
hkhaiwang@gmail.com
[2] Department of Computer Science, KAIST, Daejeon, South Korea
fhao@kaist.ac.kr
[3] Department of Computer Science, St. Francis Xavier University,
Antigonish, Canada
{chunsheng.tom.zhu,ltyang}@gmail.com
[4] Instituto de Telecomunicações,University of Beira Interior, Covilhã, Portugal
joeljr@ieee.org

Abstract. Android is a widely used operating system in mobile devices, due to that it is free, open source and easy-to-use. However, the multimedia processing ability of current android is quite limited, as the original android multimedia engine OpenCore cannot deal with lots of commonly used video (audio) formats. Recently, several approaches are proposed to enhance the multimedia processing ability and Gstreamer based method is supposed to own the best performance. However, the multimedia processing ability of current extension multimedia frameworks are still not good enough, which weakens the potential application prospect. In this paper, we provide another android multimedia framework based on Gstreamer. Extensive experiments show that our Gstreamer based framework can greatly improve the multimedia processing ability in terms of efficiency, compatibility, feasibility and universality.

Keywords: Android, Multimedia framework, OpenCore, Gstreamer.

1 Introduction

Multimedia supported green mobile networks can offer a lot of benefits for people [1]. Released by Google and supported by OHA (Open Handset Alliance), android is a widely used open source operating system for mobile devices. Apart from its operating system character, android is also a mobile software development platform which includes operating system kernel, application framework and core applications. Because it is free, open-source and easy-to use for both application developers and users, many developers and users have converted to it and it has a very bright future in the mobile market [2] [3] [4] [5]. Moreover, many multimedia terminals such as Google TV and iPad-like terminals have been popular in recent years and android OS can been modified and ported to be applied into them. This further extends the market prospect of android.

Joel J.P.C. Rodrigues et al.: (Eds.): GreeNets 2011, LNICST 51, pp. 51–62, 2012.
© Institute for Computer Sciences, Social Informatics and Telecommunications Engineering 2012

However, as the original multimedia engine OpenCore cannot deal with lots of commonly used video (audio) formats, the multimedia processing ability of android is quite limited and it cannot satisfy the various multimedia processing demand imposed by multimedia terminal devices. Recently, several approaches are put forward to enhance the multimedia processing ability of android. Specially, [6] proposes to add some audio/video coding (encoding) libraries into the OpenCore engine as plug-ins to improve the processing ability of OpenCore. [7] and [8] try to extend the Java application framework with NDK development method to perform more functions. [9] intends to employ the Gstreamer multimedia engine to supply more multimedia services for the application client. Among these extension methods, the Gstreamer based method is supposed to be the most effective, as Gstreamer is a popular multimedia engine with rich plug-ins. Whereas, the multimedia processing ability of current android multimedia frameworks are still not good enough. For example, many frameworks can only deal with specific video (audio) formats. In this paper, another design of an android multimedia framework based on Gstreamer which greatly enhances the multimedia processing ability is presented. Extensive experiments are conducted and they show that our framework obtains high efficiency, compatibility, feasibility and universality. To the best of our knowledge, our work provides essential contribution for further research regarding Gstreamer based multimedia frameworks and their commercial applications.

For the rest of this paper, section 2 briefly introduces the system architecture of android. The original android multimedia framework is discussed in section 3. Our extended Gstreamer based multimedia framework is described in section 4. Section 5 shows the experiments and gives experimental results analysis. And section 6 concludes this paper.

2   Android System Architecture

The android system consists of five layers and each layer consists of some core components. Figure 1 shows the architecture of android. From top to down, the core components are: Applications, Application Framework, Native C libraries, Android Runtime Environment (JVM), HAL (Hardware Abstract Layer), Linux Kernel.

1) Applications. Application layer consists of many core Java-based applications, such as calendar, web browser, SMS application, E-mail, etc.

2) Application Framework. Application framework consists of many components and Java classes to allow android application developers to develop various kinds of applications. By using Java language, it hides the internal implementation of system core functions and provides the developers an easy-use API. Basically, it includes Java core class and some special components of android. Some typical components are as follows: View (List, Grids), Content Provider, Resource Manager, Activity Manager.

3) Native C Libraries. In Native C library layer, it consists of many C/C++ libraries. And the core functions of android are implemented by those libraries.

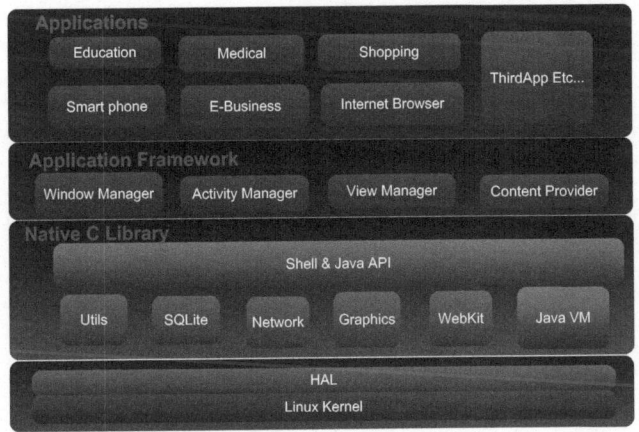

Fig. 1. Android architecture

Some typical core libraries are as follows: Bionic C lib, OpenCore, SQLite, Surface Manager, WebKit, 3D library.

4) Android Runtime Environment. Runtime environment consists of Dalvik Java virtual machine and some implementations of Java core libraries.

5) HAL. This layer abstracts different kinds of hardwares and provides an unified program interface to Native C libraries. HAL can make Android port on different platforms more easily.

6) Linux Kernel. Android's core system functions (e.g., safety management, RAM management, process management, network stack) depend on Linux kernels.

3 Android Multimedia Framework

3.1 Overall Multimedia Architecture

The android multimedia system includes multimedia applications, multimedia framework, OpenCore engine and hardware abstract for audio/video input/output devices. And the goal of the android multimedia framework is to provide a consistent interface for Java services. The multimedia framework consists of several core dynamic libraries such as libmediajni, libmedia, libmediaplayservice and so on [4].

A general multimedia framework architecture is shown in Figure 2. From Figure 2, we can see that, Java classes call the Native C library Libmedia through Java JNI (Java Native Interface). Libmedia library communicates with Media Server guard process through Android's Binder IPC (inter process communication) mechanism. Media Server process creates the corresponding multimedia service according to the Java multimedia applications. The whole communication between Libmedia and Media Server forms a Client/Server model. In Media

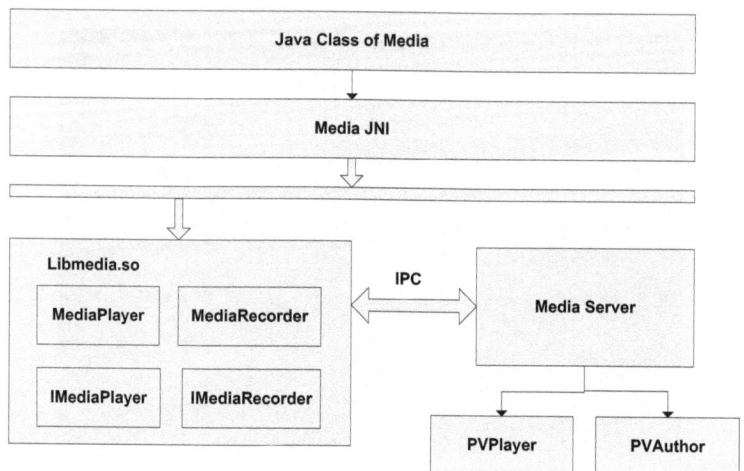

Fig. 2. Android multimedia framework architecture

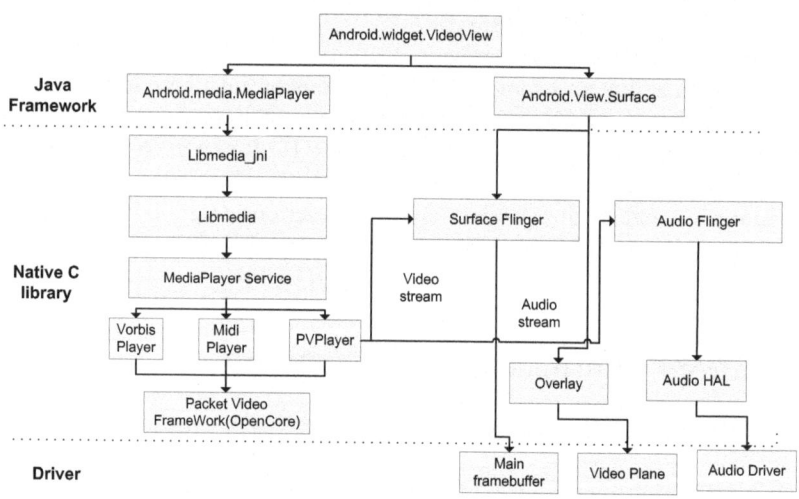

Fig. 3. Details of android multimedia framework architecture

Server guard process, it calls OpenCore multimedia engine to realize the specific multimedia processing functions. And the OpenCore engine refers to the PVPlayer and PVAuthor.

More detailed information regarding the multimedia framework are shown in Figure 3. From Figure 3, we can see that the typical video/audio data stream works in Android as follows. Specially, Java applications first set the URI of the media (from file or network) to PVPlayer through Java framework, JNI and Native C. In this process, there are no data stream flows. Then PVPlayer processes the media data stream with the following steps: demux the media data to sperate video/audio data stream, decode video/audio data, sync video/audio time, send the decoded data out.

3.2   OpenCore Multimedia Engine

OpenCore is the core of the android multimedia system. Generally speaking, it has the following characters. First, it should support most common audio formats and support stream media (RTSP/RTP). Second, it should be extended with the third-party Codecs. A general architecture of OpenCore is described in Figure 4.

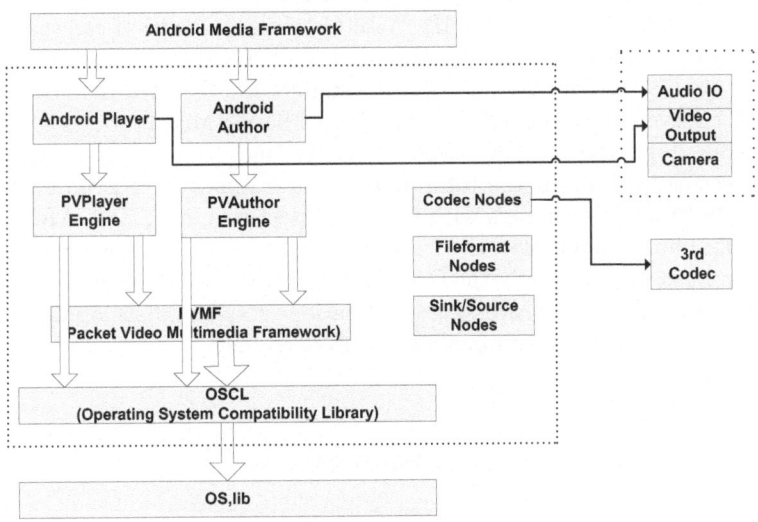

Fig. 4. OpenCore architecture

From Figure 4, we can see that OpenCore owns many functions such as media file format analysis, audio/video decoding and so on. From down to top, Open-Core includes OSCL (Operating System Compatible Layer), PVMF (Packet Video Multimedia Framework), File Formats analysis Node, Decoding Node, Encoding Node, Media I/O Node, Player engine. Furthermore, we can see that PVPlayer used in multimedia framework calls Player engine in OpenCore to realize the specific functions. At the same time, we can find that we can integrate third party Codecs into OpenCore through adding new decoding Node.

4   Gstreamer Based Multimedia Framework

4.1   Design of the Multimedia Framework

Gstreamer is a popular and widely used multimedia engine in Linux and many commercial media players use it as the core kernel [10]. The basic idea of

Gstreamer based approach is to port Gstreamer to android and it is first proposed in [9]. Compared with other extension methods (e.g., [6] [7] [8]), porting Gstreamer to android has the following advantages. First, it extends the Android's multimedia processing ability in the Native C without changing the Java API, so all the Java applications can benefit from this extension without any modification. This can avoid the limited Java application problem compared with the methods in [7] [8]. Second, Gstreamer is widely used, so there are lots of available popular plug-ins. If a new function is needed, we only need to add the corresponding plug-ins to Gstreamer. This can significantly reduces the work of extension compared with the approach proposed in [6]. Third, Gstreamer owns OpenMax IL Standard plug-in instead of using software to process video/audio. Thus, it can realize hardware video processing acceleration on the development board which supports OpenMax IL. With Gstreamer, Android can make use of the ability of hardware to the uttermost [11] [12] [13] [14].

As for our Gstreamer based multimedia framework, first, we add Gstreamer to the Media Server guard process, thus the Media Server guard process can provide multimedia play service to Java applications. Second, we change the compilation part of Gstreamer so that the total multimedia framework can run on different kinds of multimedia terminals with CPU architectures (e.g., X86, ARM, MIPS, SH4) [15]. Third, we employ the Prelink technology to enhance the execution efficiency. Last, we also apply the OpenMax IL Standard, thus our approach can support hardware video processing acceleration in special development boards. Figure 5 shows our extension method with Gstreamer.

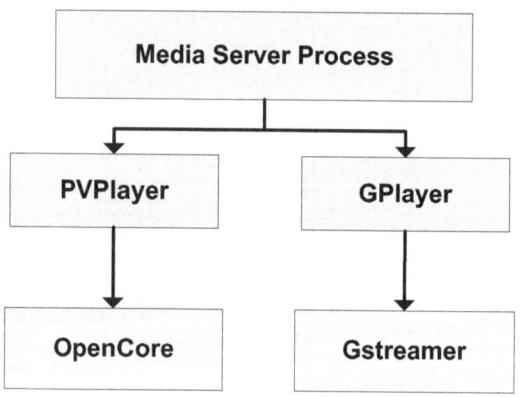

Fig. 5. Our Gstreamer-based multimedia framework architecture

To the best of our knowledge, current android multimedia frameworks exploiting Gstreamer are almost all based on the first Gstreamer based android multimedia framework in [9]. Here, different from the Gstreamer multimedia framework in [9], our framework has the following advantages. First, we still keep OpenCore instead of replacing OpenCore, as OpenCore is more sufficient than Gstreamer when dealing with H.264 encoded videos. Thus our framework can take advantage of both OpenCore and Gstreamer. Second, we modify the assembly languages in Liboil and Gstreamer engine so that our framework can run on different CPU architectures. Third, by adding more sufficient plug-ins, our framework can deal with more video (audio) formats and it has a high compatibility to different Video/Audio clips. Finally, we make some performance optimizations due to that Gstreamer has some performance weaknesses under embedded environment.

4.2   Implementation of the Multimedia Framework

To achieve the desirable functions, we need to the following work.

1) Porting some Gstreamer required open-source libraries to Android, such as Glib, Liboil, etc.

2) According to Gstreamer framework, write two plug-ins. One is used to send the decoded original video data from Gstreamer to Android display system (Surfaceflinger). And another is used to send the decoded original PCM audio data to Android audio system (Audioflinger).

3) Based on Gstreamer, we should construct a mediaplayer which can be used in Media Server guard process to supply media player service.

4) Modify assembly codes to apply Gstreamer to diversified CPU architectures, and apply some commonly used optimized technologies in embedded environment to Gstreamer.

As the porting work of Glib, Libiol and other related libraries are easy, here we introduce the rest steps.

1) The Surfaceflinger. Surfaceflinger is an important part of Android's display system. And its specific implementation mechanisms are quite complex. Moreover, it provides programming interfaces to Native C users and Java users. To send our decoded raw data to Surfaceflinger to display, in the native C user, we can do the following.

First, create a display layer through Surfaceflinger Client and get the control interface, namely, ISurface interface. Second, open the Session between Surfaceflinger and the ISurface Interface. Third, utilize ISurface to control the Layer's attributes, such as Alpha, Z-order, then close the session. Fourth, clear the previous display buffer, then post the current buffer to the bufferheap in ISurface. Last, destroy display buffer and unregister ISurface interface.

The core codes for implementation are described as follows.

2) The Audioflinger. Audioflinger is an important part of Androids audio system. Meanwhile, it provides programming interfaces to Native C users and Java users. To send our decoded raw data to Audioflinger to play, in the native C user, we need to do the following.

First, create an AudioTrack through Audioflinger Client and obtain the control interface, namely, ISurface interface. Second, open the Session between Audioflinger and the ISurface Interface. Third, utilize ISurface to control the attributes

---

Algorithm 1. Core Codes for Implementation

```
1: SurfaceComposerClient client= new SurfaceComposerClient();
2: client.Register();
3: OpenSession();
4: ISurface control=client.get();
5: control.setAlpha();
6: control.setLayer();
7: CloseSession();
8: control.clear();
9: control.BufferHeap= SurfaceSink.get().BufferHeap;
10: control.Post(BufferHeap);
11: client.UnRegister();
```

---

of the Layers, such as FrameCount, Audio Format, then close the session. Fourth, clear the previous audio buffer, then post the current buffer to the bufferheap in ISurface. Last, destroy display buffer and unregister ISurface interface.

The core codes for this part are almost the same with that of Surfaceflinger in Algorithm 1.

3) The Mediaplayer. After we build all the essential Gstreamer plug-ins, we need to use Gstreamer to construct a media player inherited from Android MediaPlayer interface to provide media processing service for Java framework. In those steps, we can use some low-level elements in Gstreamer to construct mediaplayer. But this method is extremely complex. For example, regarding a specific video clip, we need to analyze which video container format it belongs to. Then according to different formats, we need to prepare different demux, decode elements, and connect all the essential elements in a pipeline to let the whole video processing steps run in it. As different video clips correspond different processing elements, for practical use and stability, we use the Gstreamer existed high-level element "Playbin2" to start the pipeline. Then we add a bus to Playbin2 to let the video data and video processing event flow in the bus. Finally, we start the pipeline and bus. During implementation, we use gst-ffmpeg and gstopenmax plug-ins to decode video/audio data. When running our framework, the Gstreamer sends the original video data (RGB 555) to Surfaceflinger, and sends original audio data to Audioflinger. As for media record service, media meta information service and play service for H.264 encoded videos, we still use OpenCore because OpenCore can provide the whole essential functions, there is no need to extend it. Figure 6 shows the whole system after our extension.

4) Optimization methods. Due to Gstreamer is mainly used in Linux for PC, for usage in diversified embedded environment, we need to do some optimizations.

First, because Gstreamer is a plug-in based multimedia engine, the whole system needs to load many different functional dynamic libraries to coordinate to complete the whole video processing steps. Large numbers of dynamic libraries' loading and unloading can consume much memory, hence it seriously cuts down the execution efficiency. To solve this problem, we use the Prelink technology to

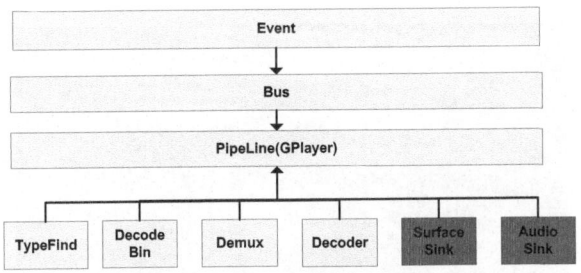

Fig. 6. The whole system after extension

give every dynamic library a fixed memory loading address to reduce the execution consumption. Second, due to that different kinds of multimedia terminal devices have different CPU architectures, we must make our extension approach work well in the common CPU architectures such as ARM, X86, MIPS and SH4. Here, we modify the assembly languages in Liboil and Gstreamer engine according to different CPU architectures. Third, to deal with more video/audio containers, we add many related Gstreamer plug-ins, such as TS Demux plug-ins. Further, we fix bugs in those plug-ins when running in embedded environment to obtain a high compatibility. Last, we add the OpenMax IL standard support in Android.

5 Experiment Results and Analysis

5.1 Experiment Setup

In our experiment, we consider several different kinds of video container formats. For each video container format, we take several different video clips with different rates, decoding standards and definitions. Though both our Android OS and Gstreamer have been ported to X86, MIPS, ARM and SH4, in this experiment section, we take S3C6410 ARM processor as an example. The development board is Real6410 as show in Figure 7. And the specific parameters are: ARM11 Samsung S2C6410 up to 667MHz, 256MB mobile DDR RAM up to 266MHz, 1GB NAND Flash, Linux 2.6.28. During the experiments, we add our Gstreamer module to Android 2.1, mount the modified Android file system to development board, then install the original Java multimedia application (without any extension) released by Google to test our work.

5.2 Experiment Results

The running results of dealing with four common video container formats (i.e., AVI, RMVB, TS, and FLV) with our Gstreamer-based multimedia framework are shown in Figure 8. Detailed performance evaluations are presented in Table 1. In the column of Table 1, 1 standards for the method using OpenCore only,

Fig. 7. Real6410 development board

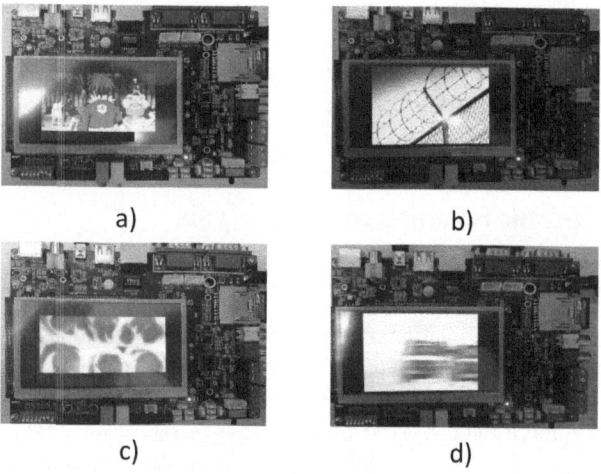

Fig. 8. Test results using AVI (a), RMVB (b), TS (c) and FLV (d)

2 represents the extended OpenCore in [6], 3 means the Java+FFMpeg+NDK approach in [7] [8], and 4 refers to our new Gstreamer based extension approach.

From Figure 8 and Table 1, we can see that, with our approach, android can deal with more common used video container formats without any change in Java applications. As for decoding efficiency, stability and compatibility, for MP4&3GP continer formats, OpenCore is more efficient than Gstreamer. As for other video formats such as ts, mpeg, etc, OpenCore cannot deal with them, Gstreamer and approach proposed in [7] (Java+NDK+FFMpeg) almost have the same efficiency. Moreover, from Table 1, we can find that using Gstreamer as multimedia engine has a high compatibility for different video formats.

Table 1. Evaluation results

VideoClips(Video+Audio)		Play	Fluent	Play	Fluent	Play	Fluent	Play	Fluent
		1		2		3		4	
TS File	MPEG2+MP2	No	Unknown	Yes	Fluent	No	Unknown	Yes	Fluent
	MPEG2+AAC	No	Unknown	No	Unknown	No	Unknown	Yes	Fluent
RMVB File	Real+Real	No	Unknown	No	Unknown	Yes	Fluent	Yes	Fluent
	WMV+AMR	No	Unknown	Yes	Fluent	Yes	Fluent	Yes	Fluent
AVI File	WMV+MP3	No	Unknown	Yes	Fluent	Yes	Fluent	Yes	Fluent
	WMV+WAV	No	Unknown	Yes	Fluent	Yes	Fluent	Yes	Fluent
FLV File	M4V+MP4	No	Unknown	Yes	Fluent	No	Unknown	Yes	Fluent
	QT+MP4	No	Unknown	No	Unknown	No	Unknown	Yes	Fluent
MP4 File	MPEG4+MP2	Yes	Very	Yes	Fluent	Yes	Fluent	Yes	Fluent
3GP File	MPEG4+MP2	Yes	Very	Yes	Fluent	Yes	Fluent	Yes	Fluent

Apart from that, our approach is more feasible. For example, if a new function is needed, we only need to add some existed plug-ins. We do not need to change the Java API and all Java applications can benefit from our extension. Furthermore, our new multimedia framework can work well on different CPU architectures with an appropriate execution efficiency, thus our extension can be applied to different multimedia terminal devices. Moreover, with OpenMax IL Standard, our approach can support hardware video processing acceleration. All these show that our approach can significantly extend the Android's multimedia processing ability regarding efficiency, compatibility, feasibility and universality.

6   Conclusions and Future Work

In this paper, focusing on enhancing the multimedia processing ability of android, the extension mechanisms regarding its core media engine OpenCore are discussed and analyzed. Paying particular attention to the Gstreamer based method, we optimize the commonly used Gstreamer based android multimedia frameworks and propose another android multimedia framework which owns good working efficiency, compatibility, feasibility and universality. As for our future work, we plan to consider some other factors such as using hardware Overlay system to accelerate video display output [16].

Acknowledgment. Part of this work has been supported by Instituto de Telecomunicações, Next Generation Networks and Applications Group (NetGNA), Portugal.

References

1. Wang, X., Vasilakos, A., Chen, M., Liu, Y., Kwon, T.: A Survey of Green Mobile Networks: Opportunities and Challenges. ACM/Springer Mobile Networks and Applications (2011)

2. Felker, D.: Android Application Development For Dummies. For Dummies, Australia (2010)
3. Meier, R.: Professional Android 2 Application Development. Wrox Press, USA (2010)
4. Conder, S.: Android Wireless Application Development. Addison-Wesley Press, Boston (2010)
5. Ableson, F.: Android in Action. Manning Publications, Greenwich (2011)
6. Song, M.Q., Sun, J., Fu, X.L., Xiong, W.K.: Design and Implementation of Media Player Based on Android. In: 6th International Conference on Wireless Communications, Networking and Mobile Computing (September 2010)
7. Song, M.Q., Sun, J., Fu, X.L.: Research on Architecture of Multimedia and Its Design Based on Android. In: International Conference on Internet Technology and Applications, pp. 1–4 (August 2010)
8. Fu, X.L., Wu, X.X., Song, M.Q., Chen, M.: Research on audio/video codec based on Android. In: 6th International Conference on Wireless Communications, Networking and Mobile Computing (September 2010)
9. Gaignard, B.: GStreamer as multimedia framework in Android: a new alternative. In: CELF Embedded Linux Conference Europe (CELF ELF Europe) (October 2010)
10. Gstreamer [EB/OL], http://gstreamer.freedesktop.org/
11. OpenMAX Integration Layer Application Programming Interface Specification Version 1.0. The Khronos Group Inc. (2005)
12. OpenMAX Development Layer Interface Specification Version 1.0.2. The Khronos Group Inc. (2005)
13. Gstreamer TI [EB/OL], https://gstreamer.ti.com/
14. Alejandro, A.R., Mireya, S.G., Sunil, K.: Streaming media portability with the emerging support OpenMAX. IETE Technical Review (Institution of Electronics and Telecommunication Engineers, India) 28, 146–157 (2011)
15. Truman, T.E.: InfoPad multimedia terminal: A portable device for wireless information access. IEEE Transactions on Computers 47, 1073–1087 (1998)
16. Lee, S.C., Jeon, J.W.: Evaluating performance of android platform using native C for embedded systems. In: Proceedings of the International Conference on Control, Automation and Systems, pp. 1160-1163 (2010)

Bandwidth Aware Application Partitioning
for Computation Offloading on Mobile Devices

Feifei Wu, Jianwei Niu, and Yuhang Gao

School of Computer Science and Engineering, Beihang University,
Xueyuan Road. 37, 100191 Beijng, China
{wufeifei,niujianwei,gaoyuhang}@buaa.edu.cn

Abstract. Computation offloading is a promising method for reducing power consumption of mobile devices by offloading computation to remote servers. For computation offloading, application partitioning is a key component. However, making a good application partitioning is challenging, as it needs to carefully consider the tradeoffs between the communication cost and computational benifits. Most of previous work makes application partitioning by using a static bandwidth to measure the communication cost and thus cannot adapt to scenarios with dynamic bandwidth. To address this problem, in this paper, we propose a Bandwidth Aware Application Partitioning Scheme (BAAP). BAAP models the bandwidth as a random variable and formulate the application partition as a 0-1 Integer Programming with Probability (IPP) problem. Then BAAP adopts Branch and Bound algorithm to solve the problem. Experimental results show that BAAP can greatly reduce energy consumption while satisfying the cost and time constraints with guaranteed confidence probabilities regardless of different network bandwidth.

Keywords: Computation offloading, Graph Partitioning, Energy Saving, Mobile Devices, Confidence Probability.

1   Introduction

With the fast development of mobile technologies, mobile devices, such as smart phones, have become the primary computing platform for many users, which can provide a range of services and applications. However, the limited battery life is still a big obstacle for the further growth of mobile devices. Various studies have identified longer battery lifetime as the most desired feature of mobile devices, therefore, prolonging the battery life of mobile devices has become one of the top challenges.

Much work has been done to address this problem by offloading computation from smart phones to remote resource rich servers [1][2][3][4] to reduce energy consumption, which is called computation offloading or code offloading. For example, Diaconescu [4] proposed a compiler and runtime infrastructure for automatically partitioning and offloading java applications, and Roelof Kemp proposed Cockoo[3], an offloading system for android applications.

Joel J.P.C. Rodrigues et al.: (Eds.): GreeNets 2011, LNICST 51, pp. 63–72, 2012.
© Institute for Computer Sciences, Social Informatics and Telecommunications Engineering 2012

The key component of computation offloading is application partitioning, which partitions the application into one local execution part and one or more remote execution parts. However, making a good application partitioning is challenging, as it needs to carefully consider the tradeoffs between the communication cost and computational cost: running a part of application locally will cause much energy consumption as the CPU needs to conduct complex computation while running the part remotely will led to extra communication cost (energy, money and delay) for transmitting the application state, code, and so on.

A traditional solution to application partitioning is to model it as a graph partitioning problem. The application is represented as a Weighted Object Relation Graph (WORG), whose nodes represent the application components and edges represent the interaction between components. The weights of nodes and edges represent the computational cost and communication cost between components respectively. Then an optimization algorithm can be applied to solve the graph partitioning problem to minimize the cost.

Apparently, an accurate measurement of communication cost is critical to partitioning decision, and the communication cost is largely decided by the network bandwidth. Some previous work [4][5][6] uses a static bandwidth to compute the communication cost when making application partitioning and thus cannot be applicable to those scenarios in which bandwidth is changing dynamically. In this paper, we propose a Bandwidth Aware Application Partitioning (BAAP) algorithm, which can adapt to dynamic bandwidth. We model the bandwidth as a random variable and construct a probabilistic WORG of the application with the weights of edges as a function of the random variable. Then, we formulate the application partitioning problem as a 0-1 Integer Programming with Probability (IPP) problem. Then BAAP adopts Branch and Bound to solve the problem to optimally determines which part of the application should be run locally and which part should be run remotely. Experimental results show that our algorithm can greatly reduce energy consumption while satisfying the cost and time constraints with guaranteed confidence probabilities regardless of different network conditions.

The rest of this paper is organized as follows: Section 2 provides a detailed description of our Bandwidth Aware Application Partitioning scheme. Experiment and analysis are presented in section 3. Section 4 provides some concluding remarks.

2   Bandwidth Aware Application Partitioning

2.1   Overview

Figure 1 shows the schematic workflow of our Bandwidth Aware Application Partitioning (BAAP) Algorithm. Currently, BAAP is target to Java Applications, but can be extended to any other Object-Oriented Applications (e.g., C++, C#). Taking a Java Application as input, BAAP first constructs the WORG of the application by static Call Graph Analysis and dynamic Profiling. Then, the constructed WOGR is passed to the Graph Partitioning module for partitioning. Based on the collected Bandwidth information and Application Specific Constraints(time constraint, cost constraint and energy constraint), the Graph Partitioning module works out the optimal assignment of each WORG node to run locally or remotely that fulfill the constrains. Once the graph is partitioned, a distributed version of application will be generated to automatically offload the code to remote server according to the partitioning results. The details of WORG constructing and Graph Partitioning will be discussed in section 2.2 and 2.3, respectively.

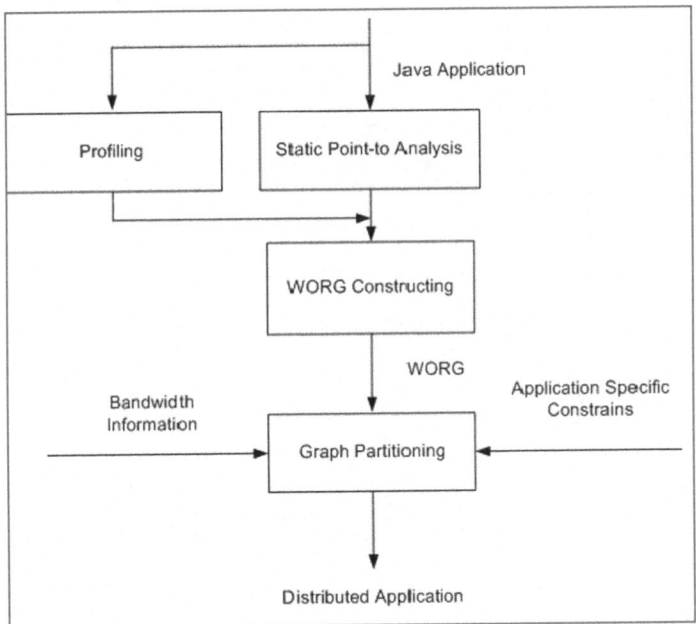

Fig. 1. Schematic workflow of BAAP

2.2   WORG Construction

BAAP models the application as a Weighted Object Relation Graph(WORG), with each node represents a run time object of the application and each edge represents the interaction between objects (we define three types of interaction: C<Creation>, I<Invocation>, D<DataAccess>). Moreover, each node is associated with a weight <CPU> to indicate the CPU execution time for each object and each edge is associated with a weight <DataCount> to indicate the total data amount that needed to be transmitted between two nodes.

In order to get an accurate WORG of the application, we first construct an initial Object Relation Graph (ORG) of the application by static point to analysis [8] and then we use offline profiling to assign weights to the ORG. Our WORG constructing algorithm is based on the method proposed in [7]. For convenience, we will use a simple example to work through our WORG constructing algorithm.

An Example. Figure 2 shows an example of java application. The example contains two main classes: an Account class describes a bank account and a Bank class describes a bank that processes those bank accounts. In addition, the Bank class creates a Vector object to save the accounts. The main function of the Bank class create one bank object and two account objects, performing operations on them through method calls. Our target is to construct a graph to represent these objects and the interaction between them.

```
class Account {
        int id;
        String name;
        int money;
        public Account(int id, String name, int money) {
                this.id = id;
                this.name = name;
                this.money = money;
        }
        public void setMoney(int money) {
                this.money = money
        }
        . . . . . . . . . . . . . . . . . . . . .
}
public class Bank {
        Vector accounts = new Vector();
        String name;
        public Bank(String name) {
                accounts = new Vector();
                this.name = name;
        }
        public void addNewAccount(Account e) {
                accounts.add(e);
        }
        public Account getAccount(int accountID) {
                . . . . . . . . . . . . . .
        }
        public void saveMoney(int accountID, int money) {
                this.getAccount(accountID).setMoney(
                this.getAccount(accountID).checkMoney()+money);
        }
        public static void main(String args[]) {
                Bank bank = new Bank("ICBC");
                Account A1 = new Account(1, "jane");
                Account A2 = new Account(2, "anne");
                bank.addNewAccount(e1);
                bank.addNewAccount(e2);
                bank.savaMoney(2, 20000);
                . . . . . . . . . . . . . . . . . . . . . . . . . . . . . . . . . . . . . . . . .
        }
}
```

Fig. 2. An Example of java application

Static Point-to Analysis. We use the Soot analysis framework [8] to perform Point-to Analysis of the application. Soot is a Java optimization framework which provides tools and APIs for analyzing and transforming java byte code. By using Soot's build-in Pointer Analysis Research Kit (Spark) and call graph analysis, we can construct an initial ORG of the example application, which is showed in Figure 3.

The OGR has five nodes and each node is annotated with S_ or D_ prefix to indicate static or dynamic objects. The entry point of the ORG is S_Bank which contains the main function. The main function first creates a Bank Object and two Account Objects which are represented by D_Bank,D_Account_1, D_Account_2 respectively. Meanwhile three edges are added from S_Bank to D_Bank, D_Account_1, D_Account_2 to indicate the creation and invocation interaction between them. In addition, the main function invokes D_Bank's "addNewAccount" method to add D_Account_1 and D_Account_2, thus two edges are added from

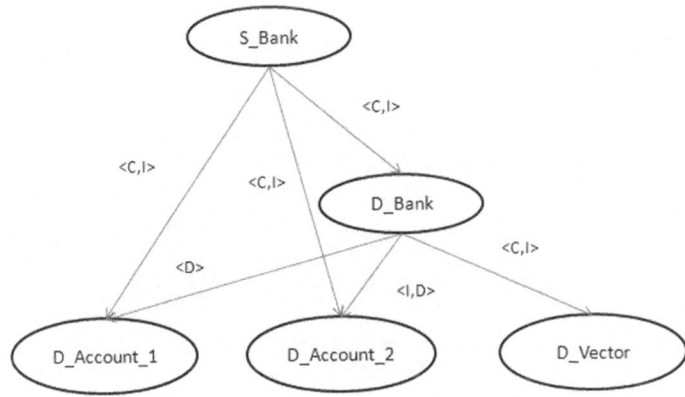

Fig. 3. Initial ORG of the Example

D_Bank to D_Account_1 and D_Accout_2 for the DataAccess interaction. Then, the main function calls the D_Bank's "saveMoney" function to set D_Account_2's "money" attribute which will in turn invocate D_Account_2's "setMoney" method, thus the edge between D_Bank and D_Account_2 is labeled with I (Invocation).

Offline Profiling. After constructing the initial ORG, we perform Offline Profiling to assign weights to the nodes and edges of the ORG to form the Weighted ORG (WORG). We assign each node with a <CPU> weight to indicate the execution time of the corresponding object, and each edge with a <DataCount> weight to indicate the total data amount that needs to be transmitted between two nodes. In order to estimate the <CPU> and <DataCount> metrics, we combine Soot's flow analysis framework with the byte code rewriting to add instrument code to collect the <CPU> and <DataCount> metrics of each node and edge.

2.3   Graph Partitioning

The task of Graph Partitioning module is to work out the optimal assignment (locally execution or remote execution) for each node of the constructed WORG. We formulate the Graph Partitioning problem as a 0-1 Integer Programming with Probability (IPP) problem and adopt Branch and Bound algorithm to solve the IPP problem. Section 2.3.1 and 2.3.2 present the details of IPP and Branch and Bound, respectively.

Problem Formulation

Input. The IPP problem takes the following inputs:
1) $G = <V, E>$: The WORG of the application, with each node Vi associated with weight <cpui> and each edge with weight <datacounti> .
2)   F(b): The Probability Distribution Function of the bandwidth b;
3) $<E_C, T_C, C_C, P_C>$: The application specific constrains, which indicates the total energy consumption, execution time and communication cost should not exceed EC, TC and CC respectively with guaranteed probability PC.

4) B: Current bandwidth.
5) SL: The set of the nodes that needs to be run locally (e.g., the objects that process user interaction).

Target. With a given input, we can compute the total energy consumption, execution time and cost of the application by the following equations:

$$Energy(G) = \sum x_i * E(i) + \sum |x_i - x_j| * E(e_{ij}) .$$ (1)

$$Time(G) = \sum_{1 \le i \le n} x_i * t_{nli} + (1 - x_i) * t_{nsi} + \sum_{1 \le i, j \le n} |x_i - x_j| * t_{ij} .$$ (2)

$$Cost(G) = \sum_{1 \le i, j \le n} |x_i - x_j| * c_{ij} .$$ (3)

x_i indicates the assignment of each node: $x_i = 1$ means local execution of the node while $x_i = 0$ means remote execution.

$E(i)$ is the energy consumption of node i when i is running locally, which can be computed through the following equation:

$$E(i) = < cpu_i > * P_{cpu} .$$ (4)

$<cpu_i>$ is the $<CPU>$ weight of node i and P_{cpu} is the power of CPU.

$E(e_{ij})$ is the energy consumption for data transmission between node i, j when they are not running together. $E(e_{ij})$ can be computed by equation (5).

$$E(ij) = < datacount_{ij} > / b * P_{wi-fi} .$$ (5)

$<datacount_{ij}>$ is the $<DataCount>$ weight of the edge that connect node i and j.

t_{nli} and t_{nsi} is the execution time of node i when running locally and remotely, which can be computed by equation (6) and (7) , respectively:

$$t_{nli} = < cpu_i > .$$ (6)

$$t_{nsi} = t_{nli} / k .$$ (7)

k indicates that the server is k times faster than local devices.

t_{ij} is the transmission time for the communication data between node i and j :

$$t(ij) = < datacount_{ij} > / b .$$ (8)

c_{ij} is the money cost for transmitting data between node i and j :

$$c(ij) = < datacount_{ij} > * c .$$ (9)

c is the money taken for transmitting 1bit data.

The task of IPP problem is to assign each node of the WORG with the optimal x_i, to make the execution time, energy consumption and cost fulfill the given constrains with given probability confidence:

$$P\{Energy(G) < E\} > P_c .$$ (10)

$$P\{Time(G)\} < T\} > P_c \cdot \tag{11}$$

$$P\{Cost(G)\} < C\} > P_c \cdot \tag{12}$$

Branch and Bound. We perform Branch and Bound algorithm to solve the IPP problem. First, in order to reduce computational complexity, we simplify the constraints defined in IPP (denoted as constraint A) problem as follows (denoted as constraint B):

$$Energy(G)_b < E \cdot \tag{13}$$

$$Time(G)_b < T \cdot \tag{14}$$

$$Cost(G)_b < C \cdot \tag{15}$$

b is the critical bandwidth that meets $P\{B>=b\}>P_c$ and $Energy(G)_b$, $Time(G)_b$, $Cost(G)_b$ represents the energy consumption, execution time and cost of a particular partitioning scheme respectively when bandwidth is b.

Constraint B is an approximate to constraint A: if a partitioning scheme can fulfill constraint B, it will fulfill constraint A with a high probability.

Then we conduct Branch and Bound on the simplified IPP. We first transform the WORG to a DAG, perform topologic sort on the DAG, and then branch form the first node in topologic sort. We use depth-first search to traverse the search tree, and for every encountered nodes, compute the $Energy(G)_b$, $Time(G)_b$, $Cost(G)_b$, compare it with E, T, C and the current minimal energy $MinEnergy$. If $Energy(G)_b$, $Time(G)_b$, $Cost(G)_b$ don't fulfill the constraints or $E > MinEnergy$, the sub tree of the node will be cut and the search back traverses to the parent node. After finishing searching, we will get the optimal partitioning that fulfills the given constrains.

3   Evaluation

This section presents the experimental results of our BAAP algorithm. We first show the drawbacks of those partitioning algorithms that using static bandwidth, then we present the performance of our BAAP algorithm and compare it with the static bandwidth based application partitioning algorithm.

3.1   Drawbacks of Static Bandwidth Based Partitioning

To show the drawbacks of static bandwidth based partitioning algorithms that using a static bandwidth to compute the communication cost, we perform the following experiment:

1)   With a given WOGR and bandwidth B(100kb/s in our experiment), work out the optimal partition scheme (denoted as A) that minimize the energy consumption of the WORG through Branch and Bound algorithm.
2)   Change the bandwidth (from 10kb/s to 100kb/s), work out the corresponding optimal partition scheme (denoted as B), and compare the energy consumption of partition scheme A and that of scheme B at different bandwidth.

We perform the experiment on three random generated graphs that simulating the WORG of actual applications. The properties of the graphs are listed in Table 1 and the experimental results are showed in Figure 4.

Table 1. Graph Properties

Parameter	Setting		Description
Graph size	Graph1	<10,43>	The number of node and edge
	Graph2	<15,103>	
	Graph3	<30,416>	
Node weight	1-720(s)		Execution time of a node
Edge weight	1-2000(kb)		Interaction data count
Cpu power	480(mw)		Cpu's power
Wi-Fi power	880(mw)		The power of wi-fi interface
cost	0.01(RMB)		Money cost for transmitting 1 bit
k	5		Speed up of server

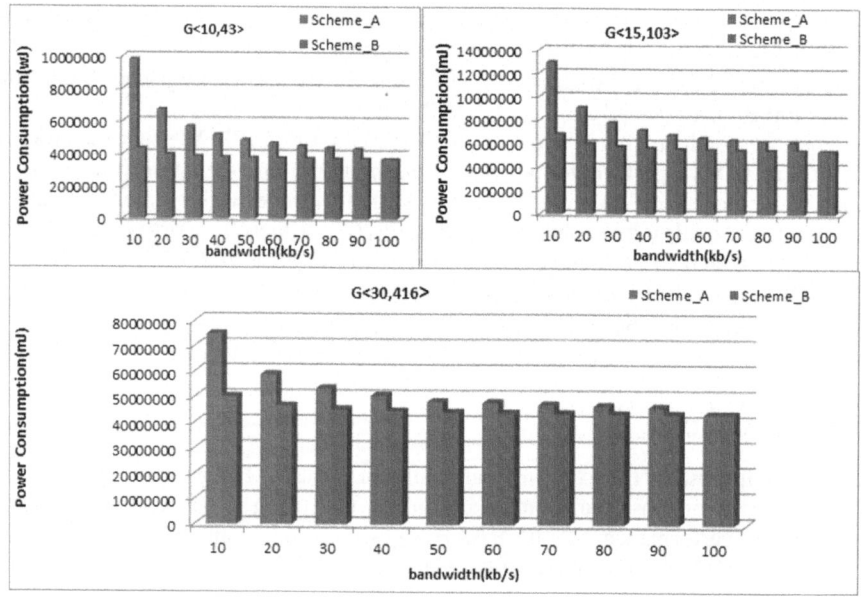

Fig. 4. Comparison of energy consumption between scheme A and B with different bandwidth

From Figure 4 we can see that with the bandwidth decreases, the energy consumption of scheme A increases and the energy consumption of scheme A is much larger than that of scheme B, especially when the bandwidth is low. The experiment results reflect that the static bandwidth based partitioning scheme cannot adapt to dynamic bandwidth: the optimal partitioning at a bandwidth may consume much more energy at another bandwidth. Therefore a bandwidth aware partitioning scheme is needed.

3.2 Performance of BAAP

To evaluate the performance of our BAAP algorithm, we work out the partitioning scheme (denoted as BAAP) through our BAAP algorithm of the three Graph that we used in 3.1 and compare the energy consumption of BAAP with that of scheme A and scheme B in 3.1 under different bandwidth(from 10kb/s to 100kb/s).

Table 2. Experiment setting

Parameter	Setting	Description
B	100	Current bandwidth
b	10	Critical bandwidth with 90% confidence
E	1.2 *E(b)	E(b) is the minimal energy with bandwidth=b
C	1.2*C(b)	C(b) is the minimal cost with bandwidth=b
T	1.2*T(b)	T(b) is the Minimal time with bandwidth=b
F(B)	B/100	B obeys the uniform distribution between 0-100

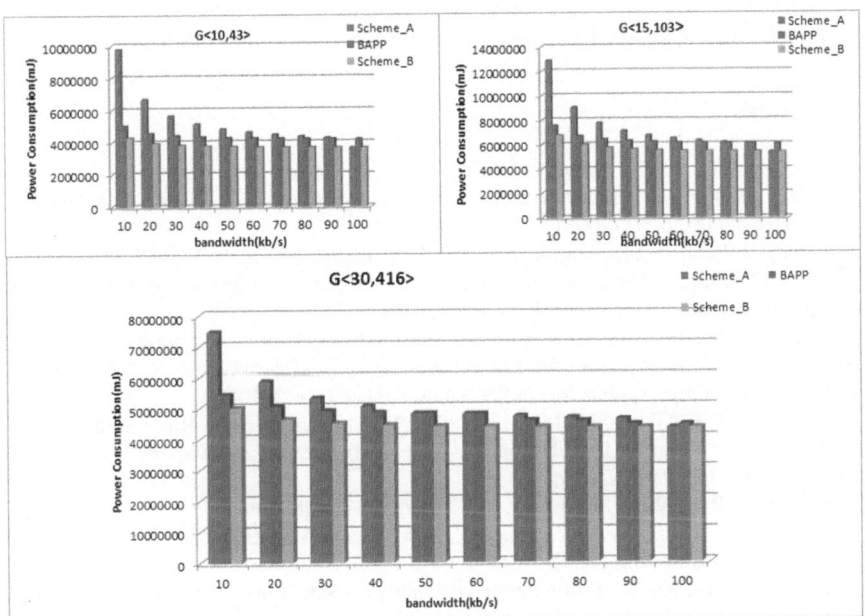

Fig. 5. Comparison of energy consumption among scheme A scheme B and BAAP with different bandwidth

Table 2 lists the constraints and parameters we set in this experiment for BAAP and Figure 5 shows the experimental results.

From figure 5, we can see that BAPP consumes less energy than Scheme_A, and thus outperforms those static bandwidth based partitioning schemes. By setting constrains to the possible partitioning schemes, BAPP can exclude those partitioning schemes that may consume much energy at low bandwidth, though they may be the optimal scheme at current bandwidth.

4   Conclusion

This paper proposed a Bandwidth Aware Application Partitioning Scheme (BAAP) for computation offloading to save energy of mobile devices. BAAP models the bandwidth as a random variable and formulates the application partition as a 0-1 Integer Programming with Probability (IPP) problem. Then, BAAP adopts Branch and Bound algorithm to solve the problem. Experimental results show our BAAP algorithm outperforms static bandwidth based partitioning schemes and can greatly reduce energy consumption while satisfying the cost and time constraints with guaranteed confidence probabilities regardless of different network bandwidth.

Acknowledgments. This work was supported by the Research Fund of the State Key Laboratory of Software Development Environment under Grant No. BUAA SKLSDE-2010ZX-13, the National Natural Science Foundation of China under Grant No. 60873241, the Fund of Aeronautics Science granted No. 20091951020, the Program for New Century Excellent Talents in University under Grant No. 291184.

References

1. Yang, K., Ou, S.: On Effective Offloading Services for Resource-Constrained Mobile Devices Running Heavier Mobile Internet Applications. IEEE Communications Magazine 46(1), 56–63 (2008)
2. Cuervo, E., Balasubramanian, A., Cho, D.: MAUI: Making Smart Phone Last Longer with Code Offloading. In: Proceedings of the 8th International Conference on Mobile Systems, Applications, and Services, San Francisco (2010)
3. Kemp, R., Palmer, N., Kielmann, T., Bal, H.: Cuckoo: a Computation Offloading Framework for Smart phones. In: Proceedings of the 2nd International ICST Conference on Mobile Computing, Application and Services, Santa Clara (2010)
4. Diaconescu, R.E., Wang, L., Mouri, Z., Chu, M.: A Compiler and Runtime Infrastructure for Automatic Program Distribution. In: 19th IEEE International Parallel and Distributed Processing Symposium, Denver (2005)
5. Diaconescu, R.E., Wang, L., Franz, M.: Automatic distribution of java byte-code based on dependence analysis. Technical Report, School of Information and Computer Science, University of California (2003)
6. Li, Z., Wang, C., Xu, R.: Computation Offloading to Save Energy on Handheld Devices: A Partition Scheme. In: Proceeding of the 4th ACM International Conference on Compilers, Architecture and Synthesis for Embedded Systems, Atlanta (2001)
7. Wang, L., Franz, M.: Automatic Partitioning of Object-Oriented Programs with Multiple Distribution Objectives. Technical Report, Donald Bren School of Information and Computer Science, University of California, Irvine (2007)
8. Soot, http://www.sable.mcgill.ca/soot/

A Content-Centric Architecture
for Green Networking in IEEE 802.11 MANETs

Marica Amadeo, Antonella Molinaro, and Giuseppe Ruggeri

University "Mediterranea" of Reggio Calabria - DIMET Department
Loc. Feo di Vito, 89100 Reggio Calabria, Italy
{marica.amadeo,antonella.molinaro,giuseppe.ruggeri}@unirc.it

Abstract. In this paper we aim to demonstrate that the emerging paradigm of content-centric networking conceived for future Internet architectures can be also beneficial from the energy efficiency point of view. The reference scenario to prove this statement is a Mobile Ad hoc Network (MANET) characterized by dynamic topology and intermittent connectivity. We design CHANET, a content-centric MANET that relies on a connectionless layer built on top of legacy IEEE 802.11 networks to provide energy-efficient content-based transport functionality without relying on the TCP/IP protocol suite.

Keywords: Content Centric Networking, Green Networking, MANETs.

1 Introduction

Mobile ad-hoc networks (MANETs) are self-organized networks of battery-powered devices that exchange information without relying on any centralized control or pre-existing network infrastructure. IEEE 802.11 [1] MANETs represent today a pervasive low-cost wireless technology thanks to the widespread diffusion of diversified 802.11-enabled handheld devices (like smartphones, tablets, MP3 players). MANET devices can actively cooperate to forward data over multihop paths towards a destination node, which can be either *any* node in the MANET or a *gateway* node offering connectivity to the Internet.

The primary usage of the current Internet as a means for discovering, uploading, accessing and sharing contents, is asking for a radical change in the underlying communication paradigm, from an *address-centric* to a *content-centric* model [2]. Several research projects are based on this idea and suggest a clean slate architecture design to build the future Internet [3] [4].

The content-centric, or information-centric, vision enables the network to focus on *what* instead of *where* data can be retrieved, through *naming* data contents instead of their location (*IP addresses*). This approach allows separating trust in data content from trust in data paths (i.e., transmission channels, hosts and servers) by naming the data through security mechanisms, with the additional advantage to enable *in-network* data caching/storing to optimize traffic management. In a content-centric network, communication is driven by receivers, which ask for a given content typically by broadcasting an *Interest* packet.

Joel J.P.C. Rodrigues et al.: (Eds.): GreeNets 2011, LNICST 51, pp. 73–87, 2012.
© Institute for Computer Sciences, Social Informatics and Telecommunications Engineering 2012

The network can satisfy the request by forwarding it to any node holding a copy of the requested content.

This decoupling in space and time between senders and receivers make content-centric networking an appealing solution also for environments with intermittent connectivity like MANETs. Recent works in the literature have shown potential beneficial effects of the content-centric paradigm in MANETs [5], [6], [7]. In this paper, we aim at designing a feasible architecture for supporting content-centric communications in a MANET; we intend to investigate whether this approach can be also beneficial from the energy efficiency point of view, and the extent of this benefit.

The proposed content-centric architecture is called CHANET (Content centric fasHion mANET). It is based on a *connectionless* content-centric layer designed on top of the IEEE 802.11 Data Link layer [1], which exploits only *broadcast* packets and *named* contents by letting each receiving node take *local* forwarding decisions based on packets overhearing.

CHANET is especially conceived to cope with dynamic topologies and intermittent connectivity with the aim of:

- keeping signalling overhead and power consumption very low;
- leveraging simplicity, availability and robustness of packet broadcasting and overhearing while keeping scalability (and broadcast storm [8]) under control by means of in-network caching and smart dissemination techniques;
- reducing power consumption by also relying on *local* forwarding decisions without explicit signalling exchange with neighbours;
- inherently supporting mobility of source and destination nodes in the MANET;
- indirectly implementing functions like error control and retransmissions, traditionally implemented at transport layer;
- providing benefits to all involved parties: to users (by allowing fast and low-power access to the requested content), and to content and network providers (by reducing operational costs of the provided sources and infrastructure).

The remainder of the paper is organized as follows. Section 2 briefly summarises related work in the area of content-centric MANETs; Section 3 describes the proposed CHANET architecture; simulation results are reported in Section 4 that show the CHANET performance against the traditional address-centric Internet model. Conclusive remarks are reported in Section 5.

2 Related Work

Content-centric network architectures proposed for future Internet [3], [4] share some common functions:

- *Content naming and security* – contents are provided with globally unique names from a flat or hierarchical space. Names are often self-certified to securely verify the authenticity of the content and the publisher.

- *Content discovery* – network nodes cooperate to forward content requests towards one or more nodes that store the content.
- *Content delivery* – it consists in forwarding the discovered content from a storage node to the subscriber. The mechanisms can exploit either route information encoded in the packets header or they keep track of each forwarded request by caching information about the interface where the request has been received from. This interface will be used to subsequently forward the retrieved data over the path back to the subscriber, like in [2].
- *In-network content caching* – network nodes can temporarily store forwarded contents, so that they can directly send the data back to the requesting node, instead of forwarding the request upstream.

Authors of [5] and [6] showed the effectiveness of a content-centric approach in a MANET. They designed a topology-agnostic data-centric forwarding protocol, named Listen First, Broadcast Later (LFBL), that exploits packet overhearing by each intermediate node to limit the drawbacks of broadcasting interests. All forwarding decisions are taken by the receiver with a minimal amount of state in each node. The data exchange phase follows the rules of *distance-based* forwarding with collision avoidance. Performance evaluation shows that LFBL significantly outperforms the traditional Ad hoc On-Demand Distance Vector (AODV) protocol [9] in highly dynamic environments.

In [7], the content-centric approach is implemented on a large scale tactical/emergency MANET with high mobility and lossy channels. Representative experiments show the superiority of the proposal over traditional proactive routing like Optimized Link State Routing Protocol (OLSR).

Energy efficiency is a critical requirement in MANETs due to the battery-powered nature of mobile device. Considerable research has been devoted in the past to low-power protocol design in an effort to enhance energy efficiency of MANETs [10]. On the other hand, in [11] and [12], the authors proved that the content-centric networking of [2] opens new possibilities for energy-efficient content dissemination in wired scenarios compared to traditional content delivery networks, where content is fetched from the origin server. Energy saving mainly comes from reducing hop counts by storing content at the intermediate nodes. Content-centric communications revealed to be energy-efficient also in sensor network environments [13] where used to manage short packets in a stationary network.

In this paper, we are interested to assess effectiveness of the content-centric approach in the dynamic MANET environment, where nodes are solicited to exchange a significant amount of data between mobile source and receivers.

3   The CHANET Architecture

As shown in Fig. 1(a), the CHANET architecture relies on a content-centric *connectionless* layer built on top of the 802.11 Data Link layer.

CHANET defines three message types for content discovery and delivery, illustrated in Fig. 2: *Interest* used to request the first content chunk, *Int-Ack* used

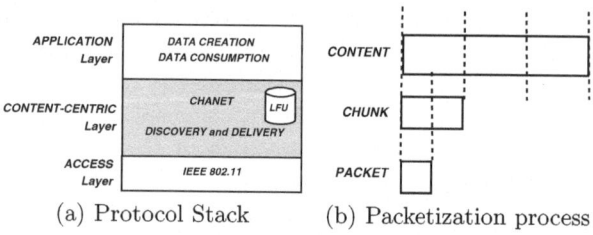

(a) Protocol Stack   (b) Packetization process

Fig. 1. CHANET protocol architecture

to request subsequent chunks and to acknowledge previously received packets, and *Data-Object (D-Object)* containing data chunks.

The CHANET communication model is based on two phases: *(i) content discovery*, in which the consumer sends the first *Interest* to search for a given content; and *(ii) content delivery*, in which *D-Object(s)* are transferred to the intended receiver while new chunk requests and acknowledgements for previous chunk(s) are sent by means of *Int-Ack* packets.

The proposed architecture is *independent* on a specific naming scheme used at the application layer, provided that the name of the searched content is passed from application to the CHANET layer. In the conceived framework, each data content (e.g., a MP3 file, a YouTube video, proximity advertising information) has its own unique and persistent name, called *Content Identificator* (*CID*). Any content is divided into *chunks*. Each chunk has its own identifier (*chunkID*) and is provided with additional control information, e.g., the digital signature for securing it. The CHANET layer implements chunk *fragmentation and reassembly*: chunks are fragmented into packets before passing through the link layer (Fig. 1(b)). The size of chunks and packets is not fixed, but it can be decided by CHANET taking into account information from lower layers (e.g., bit error rate, channel quality).

Contents sources in the MANET can be either fixed stations (e.g., an access point playing the role of a gateway to the Internet) or mobile devices. The content can be originally owned by the node (e.g., photos locally uploaded from a camera) or downloaded by a remote server and available for distribution in the MANET. Sources are allowed to periodically send *Content Advertisement* messages in the MANET to spread information about the owned or downloaded content. Fixed nodes, like access points (APs) can exploit the periodic beacon transmissions to piggyback their content advertisements, while mobile nodes can deactivate such a feature in order to save energy.

Each CHANET node maintains a *Content Store (CoS)* to cache temporarily contents, thus becoming itself a source. Usually, such nodes do not advertise the stored content and do not forward any received advertisement for energy conserving purposes. They only may send data in response to a content request. To save energy, CHANET nodes do not cache all overheard contents, but only those matching a pending *Interest*. Furthermore, due to limited memory, cache must be periodically purged: CHANET deletes from cache *the least frequently used* content. Implementation of more sophisticated policies are planned for future work. In the following, we refer to any node requesting a content as a *consumer*, and to any node that may satisfy the request (either the origin source or a node that keeps the content in its cache) as a *provider*.

In analogy to [2], each CHANET node maintains a *Pending Request Table* (PRT) for pending *Interest* and *Int-Ack* packets. In addition, also a *Content Provider Table* (CPT) is maintained by CHANET nodes, to keep the *nodeIDs* of the discovered providers and the distance to them. Nodes do not need an IP address: CHANET relies on MAC addresses as unique node identifiers. We recall that, since all communication is broadcast, 802.11 protocol cannot use retransmission mechanisms in case of packet collisions or losses. All retransmissions have to be coordinated by the CHANET layer.

3.1   Content Discovery

The content discovery phase relies on a *counter-based* broadcasting scheme for *Interest* forwarding. Counter-based schemes inhibit a node from broadcasting a packet, based on the number of packet copies already received by the node within a random access delay time. This technique is useful to reduce redundancy (and power consumption) and cope with the broadcast storm [8].

To request the first content chunk, a consumer C broadcasts an *Interest* that includes the "Chunk Name" in the form *CID/chunkID*, and waits for an answer. Answer can be either the reception of the requested *D-Object* sent by a provider, or the overhearing of the same *Interest* forwarded by a neighbouring node. If the consumer C does not detect any answer, CHANET schedules a new *Interest* transmission after a random defer time to reduce collision probability. If a *D-Object* is not received after a given number of attempts, CHANET declares the content *unreachable*.

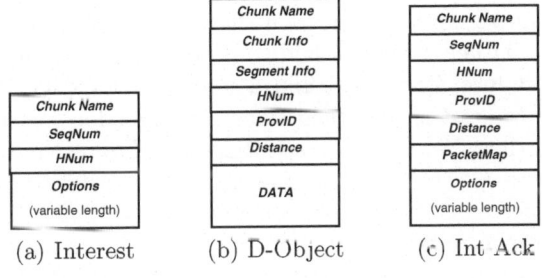

(a) Interest   (b) D-Object   (c) Int Ack

Fig. 2. CHANET packet types

As shown in Fig. 2(a), the CHANET *Interest* includes: the *Chunk Name*, a *Sequence Number* (*SeqNum*) to prevent message duplication, and a *Hop Number* (*HNum*) field that contains the number of hops the packet has crossed. *HNum* is increased by each node forwarding the packet: when it reaches its maximum value (*MaxHops*), the *Interest* is discarded. Optional fields related to the naming system can be added in order to better qualify the content that matches the *Interest* (e.g., the Publisher Public Key Digest), or to limit the area where the reply might come from (e.g., the Scope).

On the *Interest* reception, each node applies the Processing Algorithm 1.

Algorithm 1. *Interest* Processing

1: **if** $((HNum == MaxHops)$ **or** $(SeqNum$ is duplicated)) **then**
2:     Discard the *Interest*
3: **else if** (A matching is found in the CoS) **then**
4:     Compute the *D-Object Defer Time* d_d
5:     Wait(d_d)
6:     Send the *D-Object*
7:     Discard the *Interest*
8: **else if** (A matching is found in the PRT) **then**
9:     Discard the *Interest*
10: **else**
11:     Compute the *Interest Defer Time* d_i
12:     **while** (d_i is not elapsed) **do**
13:         Listen to the channel
14:         **if** (The *Interest* **or** the *D-Object* is detected) **then**
15:             Discard the *Interest*
16:             **return**
17:     Broadcast the *Interest*
18:     Insert the *Interest* in the PRT
19: **return**

More specifically, if the *HNum* field has not reached the *MaxHops* values and there is no duplication, a receiving node tries to find a match with cached content in its CoS. If a matching is found, then the node behaves like a provider P; it computes a *random defer time* d_d and waits before transmitting the *D-Object*. In case of CoS failure, it tries to find a match in its PRT. If a matching is found, then the *Interest* is discarded since a request for the same content has just been sent; otherwise, the node schedules the *Interest* re-broadcast after a *random defer time* d_i. CHANET assumes $d_d < d_i$ in order to give higher access priority to *D-Object* over *Interest* transmission.

Nonetheless the counter-based approach, some duplicated *Interests* could reach a provider P due to hidden terminal phenomena. CHANET assumes that P only replies to the first request and discards the others.

3.2 Content Delivery

The content delivery phase starts upon reception of an *Interest* by a provider P storing the requested content. The provider sends the stored *D-Object* after the *random defer time* d_d to avoid collisions with other nodes storing the data.

As shown in Fig. 2(b), a *D-Object* includes the following fields: *Chunk Name*, which is the same as for the *Interest*; *Chunk Info*, which contains security and temporal information related to the transferred chunk; *Segment Info*, used for chunk reassembly at the consumer side, since the chunk may be fragmented in more *D-Object(s)* before passing to the link layer; *ProvID*, which contains the provider's *nodeID*; *HNum*, which contains the number of hops the packet crosses (as for the *Interest*); *Distance*, which contains the hop distance between P and C, obtained from the *Interest's HNum* field.

On *D-Object* reception, any intermediate node that maintains the related *Interest* in its PRT has to cancel it, to cache the data in its CoS, and to insert the newly discovered provider identifier (*ProvID*) in the CPT, including the distance to it, and the owned Content Name (identified by the *CID/ChunkID*). If the CPT entry already exists, the intermediate node only updates it. CPT entries are refreshed after any *D-Object* reception, otherwise they are purged when a *timeout* expires. Finally, the node rebroadcasts the *D-Object* after a random defer time and by using the *counter-based* approach, in analogy to the *Interest* forwarding. Nodes without a related PRT entry simply discard the packet.

By following the chain of PRT entries, the *D-Object(s)* is (are) forwarded to consumer C. If more providers have returned the data, consumer C inserts the providers' *nodeIDs* in its *CPT*; it selects the provider that may give the best performance and sets it as "preferred provider" (*PProv*). At this time, the selection algorithm depends only on the hop-count metric, thus C selects the *nearest* provider. More sophisticated choices could be implemented in the future.

After the first chunk reception and the provider selection, successive chunk requests will be broadcasted in *Int-Ack* packets, whose format is shown in Fig. 2(c). Compared to the *Interest*, the *Int-Ack* packet has three more fields: *ProvID* is the identifier of the *PProv* selected by consumer C; *Distance* represents the expected hop number between C and *PProv* (as read by the *HNum* field of previously received *D-Object*); *PacketMap* is used by C to acknowledge packets of the previously received chunk(s) so that the corrupted or lost ones may be retransmitted. *PacketMap* is a matrix whose generic element p_{ij} represents packet j (with j ranging from 1 to the number of packets in a chunk, p_{num}) in chunk i (with i ranging from 1 to a given number of recently received chunks, c_{num}). A value *1* in bit p_{ij} indicates that packet j in chunk i has been received correctly.

The *Int-Ack* may therefore carries two different content requests: the request for a new content chunk and the request for *D-Object(s)* associated to previous chunk(s) that has (have) to be retransmitted. It may happen that a receiving node has in the CoS the *D-Object(s)* to be retransmitted but not the new content chunk. In this case, we refer to *Partial CoS Matching*, while a *Total CoS Matching* happens when the node can satisfy both requests.

As for the *Interest*, *Int-Ack* processing is based on the *defer time* calculation and the *counter-based* forwarding, but three new features are introduced to improve performance in many aspects, including energy efficiency and scalability: *Int-Ack Aggregation*, *Provider Handoff* and *Selective Response*. Specifically, at any forwarding node F, if the *Int-Ack* packet is still valid, a *Total CoS Matching* is first searched. If the content matches and the node is also the *PProv*, it checks if the same request has been just satisfied for other users. This check, referred as *Int-Ack Aggregation*, allows F to discard redundant requests from neighbouring nodes asking for the same content. If the *Int-Ack Aggregation* check fails, then F immediately sends the *D-Object*. Otherwise, if the forwarding node is not the *PProv*, it considers its distance to consumer C (by reading the *HNum* field in the *Int-Ack* packet) and compares it with the *Distance* field value. If F is closer to C than *PProv*, then it schedules the transmission of the *D-Object* after a defer time d_d. During the waiting time, a counter-based algorithm is run by overhearing *D-Object* packets that could be sent. This strategy, which we call *Provider Handoff*, mainly help to cope with highly dynamic topologies: due to the node mobility, a new provider can join the MANET and offer a better service than the current preferred provider; conversely, the consumer C can move away from the current *PProv* and enter the transmission range of a new provider.

If the *Total CoS Matching* fails but a *Partial CoS Matching* exists and F is closer to C than *PProv*, F may apply the *Selective Response* routine. It consists of sending the *D-Object(s)* that the *PacketMap* requires to retransmit and then forwarding a modified *Int-Ack*, where the *PacketMap* is purged of the just transmitted packets. As in the previous cases, a counter-based algorithm is run before transmission.

If there is a matching neither in the CoS nor in the PRT, F checks whether it is on the path between C and *PProv* by looking in its CPT. If a matching is found, a counter-based algorithm decides if forwarding the packet or not. The complete *Int-Ack* processing procedure is shown in Algorithm 2.

4   Performance Evaluation

The CHANET architecture has been implemented in Network Simulator 2 (ns-2) [14] for performance evaluation. The reference scenario is illustrated in Fig. 3. We consider one origin content source fixed in the centre of a square grid of side $500m$. The source can be co-located in an AP, which represents the only infrastructure made available by the network operator to serve MANET customers in the covered area. Each user moves according to the Truncated Levy Walk mobility model [15] with a minimum speed of $1m/s$ and maximum of $1.5m/s$. We further considered a Ricean fading model, which accounts for multipath effects due to obstacles, trees and buildings.

We focus on content download to evaluate two main aspects: *(i)* the capability of CHANET to deliver the desired content to the requesting users, and *(ii)* the energetic cost, expressed in terms of Joules spent for each bit successfully delivered to the consumers. We suppose that a subset of users, randomly selected among the 25 mobile nodes in the simulated grid, are interested in downloading a set of contents provided by the AP.

We examine three cases, depending on the number of different contents downloaded from consumers: *(i) 1 CID*: all the consumers download the same

Algorithm 2. *Int-Ack* Processing

```
1: if ((HNum == MaxHops) or (SeqNum is duplicated)) then
2:    Discard the Int-Ack
3: else if (Total CoS Matching) then
4:    if (I'm the PProv) then
5:       if (Int-Ack Aggregation Check) \\Int-Ack Aggregation then
6:          Discard the Int-Ack
7:       else
8:          Send the content
9:    else
10:       if (Distance Check) then
11:          Compute the D-Object Defer Time dd
12:          if (Counter-based Check) then
13:             Discard the Int-Ack
14:          else
15:             Send the content; \\Provider Handoff
16:       else
17:          Discard the Int-Ack
18: else if ((Partial CoS Matching) and (Distance Check)) then
19:    Compute the D-Object Defer Time dd
20:    if (Counter-based Check) then
21:       Discard the Int-Ack
22:    else
23:       Send the D-Object(s) \\Selective Response
24:       Broadcast the modified Int-Ack
25:       Insert the modified Int-Ack in the PRT
26: else if (A matching is found in the PRT) then
27:    Discard the Int-Ack
28: else
29:    if (There is a CTP entry for the PProv) then
30:       Compute the Interest Defer Time di
31:       if (Counter-based Check) then
32:          Discard the Int-Ack
33:       else
34:          Broadcast the Int-Ack
35:          Insert the Int-Ack in the PRT
36:    else
37:       Discard the Int-Ack
38: return
```

content; *(ii)* *2 CIDs*: two different contents are downloaded, the first one is requested by 66,66% of consumers and the second one by 33,33% of consumers; *(iii)* *3 CIDs*: three different contents are downloaded, each one is requested by 33,33% of consumers. We compare the performance obtained with CHANET against that achievable by using an FTP connection over the traditional TCP/IP stack and when AODV [9] is used as the routing protocol in the MANET. To achieve fair comparison, TCP Vegas has been implemented because of its higher

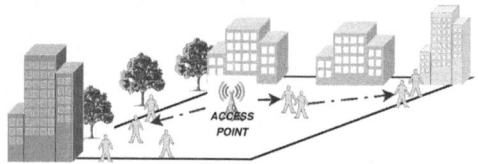

Fig. 3. Simulation scenario

performance in MANETs compared with other TCP versions [16]. It detects incipient congestion by monitoring variations of the packet delay (instead of losses).

While in the FTP over TCP scenario contents can be downloaded only through the AP, in CHANET content chunks can be also obtained by any node storing them. In both cases, we consider IEEE 802.11g as the access technology. Main parameters are reported in Table 1 as taken from the standard and from datasheets of commercially available devices [17]. Transmission power and receive sensitivity values of the simulated devices are used as inputs to the ns-2 energy model. At the beginning of the simulation, each node has assigned an initial energy level that is decremented at any packet transmission, reception and overhearing. We set the initial energy very high to be sure that no node runs out energy during the simulation.

Parameters concerning CHANET and the legacy protocol stacks are summarized in Table 2.

Table 1. 802.11g Simulation Parameters

PHY Parameter	Value
Frequency	2.4 GHz
Receive Sensitivity	-86 dBm
Transmission Power	18 dBm
Power consumption while transmitting	1.74 W
Power consumption while receiving	0.9306 W
Power consumption while idle	0.6699 W
MAC Parameter	**Value**
SlotTime	9 μs
SIFS	10 μs
Preamble Length	96 bit
CWmin - CWmax	15 - 1023
Short - Long Retry Limit	3 - 7
MAC header	34 bytes

We evaluate both network performance metrics and energy cost metrics. Concerning the former, we consider the *Download Time* as a good indicator of the user experience. It is defined as the average time required for a user to download the requested content, which in our case is assumed to be a 10MB file. Also signalling overhead is taken into account in the network performance. Specifically, we compute two types of overhead, respectively related to the consumer and to the overall network. *Consumer Overhead* is defined as the average ratio between

Table 2. Architectures' Simulation Parameters

CHANET	Value	TCP/IP/AODV	Value
Interest size	24 bytes	**TCP Vegas header**	20 bytes
Int-Ack size	40 bytes	**TCP Vegas** α	1 packet
D-Object **header** size	40 bytes	**TCP Vegas** β	3 packets
D-Object pay-load size	1000 bytes	**TCP Vegas** γ	1 packet
Chunk size	10 Kbytes	**Payload size**	1000 bytes
Aggregation time	20ms	**AODV RREQ size**	48 bytes
MaxHops	10	**AODV RREP size**	44 bytes
Defer time (Data)	$[SlotTime,\ CWmin*SlotTime]$	**AODV RERR size**	32 bytes
Defer time (Interest, Int-Ack)	$[CWmin*SlotTime,\ 2*CWmin*SlotTime]$	**AODV HELLO**	disabled

the transmitted signalling bytes and the received data bytes for a consumer. It represents a measure of efficiency from the consumer point of view. In the computation, only the signalling originated by consumers is considered, packet duplications in the network are not included. CHANET signalling packets include *Interest* and *Int-Ack*; no further overhead is taken into account, since at the MAC layer no retransmission is allowed due to the broadcast nature of all exchanged packets. For the TCP/IP case, the signalling overhead includes TCP control packets (three-way handshaking segments and all ACKs transmitted by consumers to the AP), AODV control packets (route request, route reply and route error packets), and the MAC acknowledgments for unicast data transmission (Request-To-Send/Clear-To-Send exchange is disabled). The *Network Multiplication Factor* aims to quantify the percentage of bytes totally generated into the network per each received data byte. It is defined as the ratio between all the bytes (signalling and data) sent by all nodes (source, consumers and forwarders) over the MANET and the data bytes received by all consumers. It gives a measure of the "multiplication" factor of the network since it takes into account packet duplications and retransmissions. Concerning the energy cost, we compute it in terms of *Energy per Bit*, which is defined as the Joules spent for each bit successfully delivered to the consumers. Energy consumption is evaluated for all the involved parties: network, consumers, and network operator. The *Network Energy Cost* is an overall cost parameter defined as the energy consumed by the whole network (source, consumers and all involved nodes in the MANET) to deliver the total amount of required bits to the consumers. The *Consumer Energy Cost* is the mean energy spent by a consumer to receive a bit. The *Operator Energy Cost* is defined as the energy spent by the source for each bit delivered to a consumer. When the source is an AP managed by a network operator, this energetic cost has a direct impact on the operational expenditures (OPEX).

Figure 4 shows the performance in terms of Download Time. We observe that CHANET outperforms the legacy approach thus achieving faster downloads. The delay increases with the number of consumers due to higher load and congestion, but in CHANET this trend is smoother than for traditional FTP. Motivations

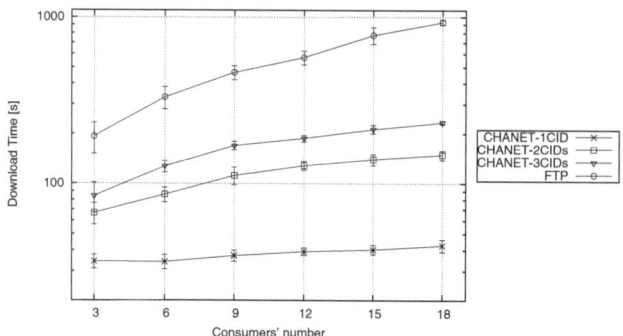

Fig. 4. Download Time

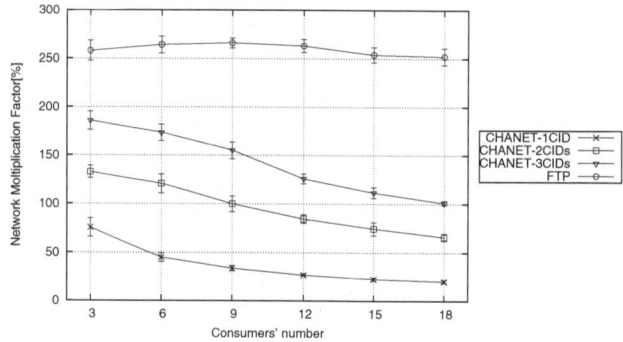

Fig. 5. Network Multiplication Factor

of such an advantage may be found in both Figures 5 and 6. In Fig. 5 the
Network Multiplication Factor is plotted against the number of consumers. With
CHANET, the network load does not increase with the consumer number, like in
the legacy suite case, rather it decreases, since a single transmission may deliver
data to more users simultaneously. This effect is amplified with the number of
receivers. The *Consumer Overhead* is also very low compared to FTP, as shown
in Fig. 6. Of course, CHANET performance get worse with the increasing of the
number of CIDs, since the probability that a single packet transmission serves
more than one user decreases.

Figures 7 - 9 summarize the protocol behaviours with respect to the energy
efficiency. CHANET is significantly better performing than the legacy case. Under some circumstances (i.e., a single CID case) the difference between the two
cases is of an order of magnitude. Figure 7 reports the Network Energy Cost
that is the most practical indicator of the pollution produced by transmissions.
Not only CHANET is more efficient than TCP/IP, but its efficiency also increases with the number of consumers. The reason is again due to the capability
of CHANET to serve more users in a single transmission, thus reducing the total number of transmissions in the network (see Fig. 5) and avoiding unnecessary

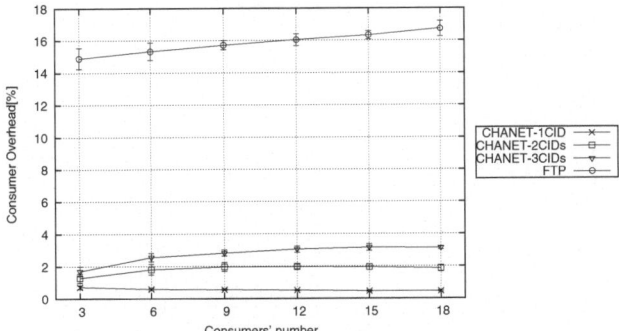

Fig. 6. Consumer Overhead

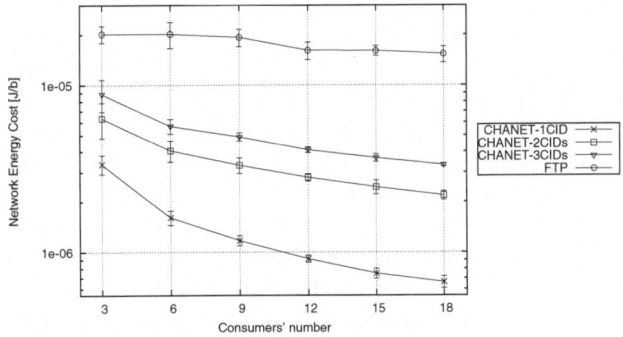

Fig. 7. Network Energy Cost

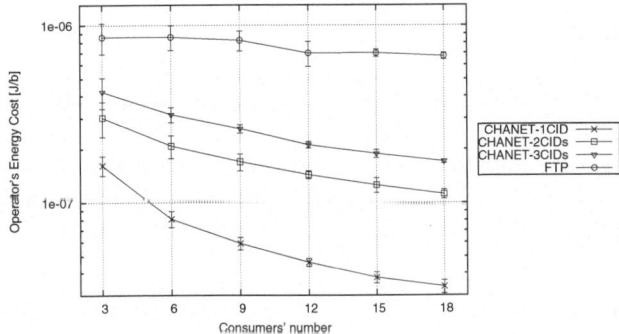

Fig. 8. Operator Energy Cost

energy consumption. Once again the efficiency worsens with the increasing number of CIDs as a direct consequence of the network load. The Operator Energy Cost represented in Fig. 8 shows that CHANET outperforms the legacy suite by ensuring to the network operator a much higher efficiency. As in the previous case, the efficiency increases with the number of consumers. From an economical point,

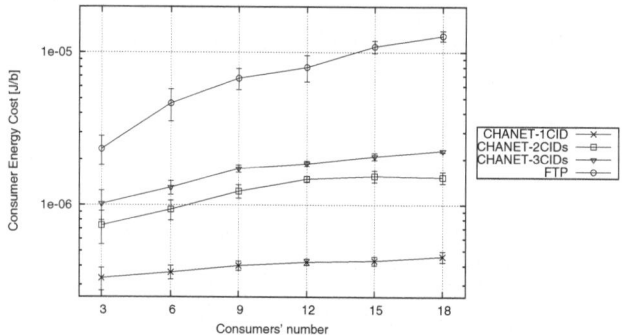

Fig. 9. Consumer Energy Cost

we may say that an operator who decides to run CHANET on its APs may get a substantial reduction in the OPEX cost. Finally, Fig. 9 reports the *Consumer Energy Cost*. Also in this case, CHANET performs better than TCP, but we observe a different trend when varying the number of CIDs. When a single CID is downloaded by all the consumers, the energy cost is lower. This is due to the packet overhearing, which allows a consumer to receive data without even requesting it and, hence, without spending energy to request it. However, overhearing becomes less effective at the increasing of the number of CIDs, since the effects of packets collisions and retransmissions prevail. By considering, for instance, the case of 2 CIDs, overhearing is not effective from 3 to 9 consumers and effects of collisions prevail so energy costs increase, while, for 12 to 18 consumers, overhearing starts to be effective again and overcomes the adverse effects of collisions.

5   Conclusions

In this paper, we developed a new content-centric energy-efficient architecture named CHANET that achieves content retrieval, delivery and caching in IEEE 802.11 MANETs. Simulation results show the great benefits offered by CHANET in terms of higher energy efficiency, reduced latency and control overhead compared to traditional MANETs based on the TCP/IP suite.

References

1. IEEE Standard for Wireless LAN Medium Access Control (MAC) and Physical Layer (PHY) Specifications, Std. 802.11-2007 (June 2007)
2. Jacobson, V., et al.: Networking Named Content. In: ACM CoNEXT, Rome, Italy (December 2009)
3. Pan, J., Paul, S., Jain, R.: A Survey of the Research on Future Internet Architectures. Communications Magazine 49(7), 26–36 (2011)
4. Ahlgren, B., et al.: A Survey of Information-Centric Networking. Dagstuhl Seminar (February 2011)

5. Meisel, M., Pappas, V., Zhang, L.: Ad Hoc Networking Via Named Data. In: ACM MobiArch 2010, Chicago, Illinois (September 2010)
6. Meisel, M., Pappas, V., Zhang, L.: Listen First, Broadcast Later: Topology-Agnostic Forwarding Under High Dynamics. In: Annual Conference of International Technology Alliance in Network and Information Science, London, UK (September 2010)
7. Oh, S.Y., Lau, D., Gerla, M.: Content Centric Networking in Tactical and Emergency Manets. In: IFIP Wireless Days, Venice, Italy, pp. 1–5 (October 2010)
8. Tonguz, O., et al.: On the Broadcast Storm Problem in Ad Hoc Wireless Networks. In: Broadband Communications, Networks, and Systems (BROADNETS), San Jose, CA (October 2006)
9. Perkins, C.E., Belding-Royer, E.M., Das, S.: Ad Hoc on Demand Distance Vector (AODV) routing. IETF, RFC 3561 (July 2003)
10. Jones, C.E., et al.: A Survey of Energy Efficient Network Protocols for Wireless Networks. Wireless Networks 7(4), 343–358 (2001)
11. Lee, U., Rimac, I., Hilt, V.: Greening the Internet with Content-Centric Networking. In: e-Energy 2010. University of Passau, Germany (2010)
12. Lee, U., Rimac, I., Kilper, D., Hilt, V.: Toward Energy-Efficient Content Dissemination. IEEE Network 25(2), 14–19 (2011)
13. Intanagonwiwat, C., et al.: Directed Diffusion for Wireless Sensor Networking. IEEE/ACM Transactions on Networking (TON) 11(1) (February 2003)
14. The Network Simulator-2 (ns-2), http://www.isi.edu/nsnam/ns
15. Rhee, I., et al.: On the Levy-Walk Nature of Human Mobility. In: IEEE INFOCOM, Phoenix, AZ (April 2008)
16. Papanastasiou, S., Ould-Khaoua, M.: Exploring the Performance of TCP Vegas in Mobile Ad Hoc Networks. International Journal of Communications Systems 17(2), 163–177 (2004)
17. Cisco aironet 802.11a/b/g wireless cardbus adapter. Data Sheet available on line at http://www.cisco.com/en/US/prod/collateral/wireless/ps6442/ps4555/ps5818/product_data_sheet09186a00801ebc29.pdf

TOA Ranging Using Real Time Application Interface (RTAI) in IEEE 802.11 Networks

Jian Fang[1], Alvin Lim, and Qing Yang[2]

[1] Computer Science and Software Engineering,
Auburn University, Auburn AL 36849 USA
{fangjia,limalvi}@auburn.edu
[2] Department of Computer Science
Montana State University, Bozeman MT 59717 USA
qing.yang@cs.montana.edu

Abstract. Ranging and positioning of wireless mobile devices using time-of-arrivals (TOA) method is becoming an increasingly interesting and challenging research topic. There are various TOA-based ranging algorithms and positioning systems, but most of them require either specially designed hardware or modifications to existing firmware. Using only off-the-self hardware, we present a novel software-based TOA ranging approach which accurately measures TOAs using the Real Time Application Interface (RTAI) operating system. A prototype system is implemented which provides precise measurements of round trip time (RTT) using IEEE 802.11b MAC layer ACK frames and the real-time communication mechanism provided by RTAI. Experiments show that using RTAI can achieve a ranging result with precision close to the accuracy obtained by hardware based methods.

Keywords: IEEE 802.11, Ranging, Localization, RTT, TOA, RTAI.

1 Introduction

In today's fast-paced and technology-centric world, positioning and tracking through mobile devices is extremely useful for applications such as disaster rescue missions, travel guidance systems, fire fighting, and healthcare services. Although GPS positioning system is widely used, it does not work correctly in indoor or metropolitan areas where tall buildings may block GPS signals. On the other hand, Wi-Fi networks are designed mainly for indoor use, so an alternative solution is localizing mobile devices through widely populated IEEE 802.11 wireless networks. In this article, we are interested in the time-of-arrival (TOA) based ranging technique which is the foundational component of IEEE 802.11 localization system.

For TOA-based ranging techniques, accurate round trip time (RTT) measurement is the most important problem. However, it is not an easy task to obtain accurate RTTs because many factors may contribute to errors in RTT measurements such as multipath effect, signal interference, operating system scheduling,

Joel J.P.C. Rodrigues et al.: (Eds.): GreeNets 2011, LNICST 51, pp. 88–98, 2012.
© Institute for Computer Sciences, Social Informatics and Telecommunications Engineering 2012

and CPU clock throttling. All these factors could introduce variations or jitters to the RTT measurement procedure. This implies that the longer the execution path that the data travels in a system, the more difficult it is to obtain a stable RTT, and the higher the errors that are introduced to the ranging results.

To address this issue, a TOA ranging system needs to measure the RTT as close to the system's bottom (physical) layer as possible. If time stamping cannot be provided at the bottom layers, a deterministic execution of instructions along the data path needs to be guaranteed, which is usually considered a difficult task due to hardware and software latency and jitters [10]. Another important factor affecting the accuracy of RTT measurements is the resolution of system timer. For instance, the programmable interval timer (PIT) on most Intel x86 CPUs has a coarse resolution of $1ms$ (1000Hz) which corresponds to a computed distance of more than $30km$. Therefore, to achieve a viable TOA ranging system, three key factors must be taken into account: 1) short data path, 2) deterministic execution of instructions, and 3) high-resolution timer.

To satisfy the three requirements for accurate RTT measurements, Real Time Application Interface (RTAI) [1] is selected in our implementation. By taking full control of hardware and task scheduling, RTAI can achieve a much faster response to outside events, e.g., interrupts of network activities, than general purpose operating systems. Therefore, more accurate time stamps can be recorded when a packet is sent or received by network devices, which is essential for obtaining precise RTT measurements.

There are mainly four contributions of this paper. First, the variation of RTT measurements caused by the processing delays at system and process levels is completely investigated. Second, we design and implement an accurate TOA based ranging system using RTAI for standard IEEE 802.11 networks. Third, a prototype system is implemented and tested using off-the-shelf wireless cards. From experimental results, we find the proposed system can achieve a precise ranging between two wireless devices with an error $< 5m$ in indoor and an error $< 8m$ in outdoor scenarios. Fourth, since the RTAI based ranging system does not require specific hardware, it can be easily installed (as kernel modules) on existing IEEE 802.11 network devices. In fact, the same methodology can be utilized in other wireless networks, e.g., sensor network and vehicular networks.

The rest of the paper is organized as follows: Section 2 presents methods that are related to our work. Section 3 analyzes the task latency at the system and process levels, and then describes the RTT ranging method in detail. Section 4 and 5 give the setup of our experiments and result analysis, respectively. Finally, conclusions are provided in Section 7.

2   Related Works

Since different applications may require different location accuracy or positioning precision, various wireless ranging methods are used by positioning and tracking systems. These techniques can be classified along different dimensions, such as hardware or software-based, time or signal based [2,3,4]. Besides the widely used

GPS system which does not work well for indoor environments, there exist several other positioning techniques such as systems based on infrared, ultrasonic, or received signal strength (RSS) [2]. Although they can overcome the shortcoming of GPS systems, these techniques are not widely adopted either because of their reliance on special hardware, offline trained radio maps, or complex design.

Another group of ranging methods are time-based techniques such as the time of arrival (TOA) and time difference of arrival (TDOA) ranging. TDOA ranging can provide accurate results but require perfect synchronization of APs, which adds a great amount of complexity to the positioning system. TOA is similar to TDOA but does not require synchronization. Synchronization error between two nodes can be eliminated by measuring the RTTs of network packets. For example, the RTS/CTS (request-to-send and clear-to-send) packets in MAC layer are used to measure RTTs in [5]. Using phase matching and shifting of received signals, [6] reports a ranging precision of less than $5m$. In [7], a debug version of Intel ABG WLAN card and an external FPGA card are used for accurately time stamping transmit and receive signals. All the above-mentioned systems achieve high precision ranging results but require auxiliary hardware or modifications to existing wireless devices' firmware.

Unlike those hardware-based method, a software based TOA ranging method is presented in [8]. It achieves an indoor ranging error of a few meters by modifying the driver of existing wireless devices. Our work is different from [8] because we use a RTAI extended system which largely minimizes the time variations in RTT measurements. Moreover, in the proposed system, the MAC layer's status information can be easily obtained from the application's kernel modules or user space using services provided by RTAI. Thus, it is easier to develop real-time localization systems with our approach.

3   Design and Implementation

RTT of a packet can be measured on different levels, e.g., hardware, firmware, driver, OS and application. Among these available approaches, measuring TOA in the software layer may be the most feasible choice because software can be more easily updated than hardware.

3.1   Latency at System Level

Time stamping a packet can be done on different levels, e.g., in the user space or at the driver layer. However, theoretically, a precise and accurate ranging can only be achieved by measuring the time when a data packet leaves or enters the transmitter/air boundary. Unfortunately, it is impossible to accomplish this task by a pure software based approach. For instance, the SoftTOA [8] measure RTT of network packets in the network driver layer which is above the transmitter/air boundary. That means the obtained RTT between the sender and receiver is:

$$RTT = t_{tx\_proc} + 2 \times TOA + t_{rx\_proc} \qquad (1)$$

where t_{tx_proc} is the time for the sender to transmit a packet and process the corresponding ACK message. t_{rx_proc} is the time for the receiver to process the received data packet, generate and transmit an ACK message. Equation 1 implies the ranging precision of RTT depends on the precise measurements of t_{tx_proc} and t_{rx_proc}, which are determined by the hardware and software used in the ranging system.

The hardware and software of a system affect the measurements of t_{tx_proc} and t_{rx_proc} in two ways: latency and jitter. Generally, hardware latency determines the system response time but may not affect the precision of RTT, e.g., a precise RTT can be obtained by hardware with longer but constant latency. However, jitter of latency is detrimental to precise RTT measurement. Thus, TOA ranging techniques need to focus on reducing jitters.

Generally, time jitters in the hardware are small, e.g., the time is relatively stable for a NIC (network interface controller) transmitting a fixed number of bits. However, large jitters exist in software due to the internal non-deterministic execution of instructions in a general-purpose OS. For instance, the RTT measurement process may be preempted by other process with a higher priority. To eliminate these jitters, a software-based ranging method needs to timestamp the data as close to the physical layer as possible.

3.2   Latency at Process Level

To obtain a precise RTT measurement, it is important to identify the points (in a process) where time stamps should be put. As shown in Fig. 1, a typical process (e.g., the RTT measuring process) is initiated within the user space. Then, to transmit a packet, it may enter the kernel space which may be later interrupted by a hardware signal. Obviously, measuring RTT at the interrupt level will give the best measurement accuracy because it is the level closest to the hardware.

In IEEE 802.11 networks, when a device receives a packet, the MAC layer of its NIC will generate an interrupt which notifies the OS that a packet matching its MAC address is received. This time instance in which the interrupt is generated should be a perfect time stamping point since it is the closest to the physical layer. However, the CPU in a non-real-time OS may not be able to respond to this interrupt immediately, so an accurate time stamping cannot be achieved.

For example, Task 1 in Fig. 1 may be executing in a critical region where an interrupt (of receiving a packet) is generated at time t_3. The interrupt can be responded only after t_4 or after Task 1 exits the critical region. In other words, a ranging process can timestamp this event of receiving a packet only after t_4. $t_4 - t_3$ is the latency for interrupt handling and it will introduce jitters into the measured RTT. Even worse, the time length of $t_4 - t_3$ is non-deterministic, i.e., it depends on the running process, the scheduler, and the resolution of the timer. Therefore, at the process level, a precise ranging system requires a fully preemptable kernel if time stamping needs to be provided in the upper layers of the system.

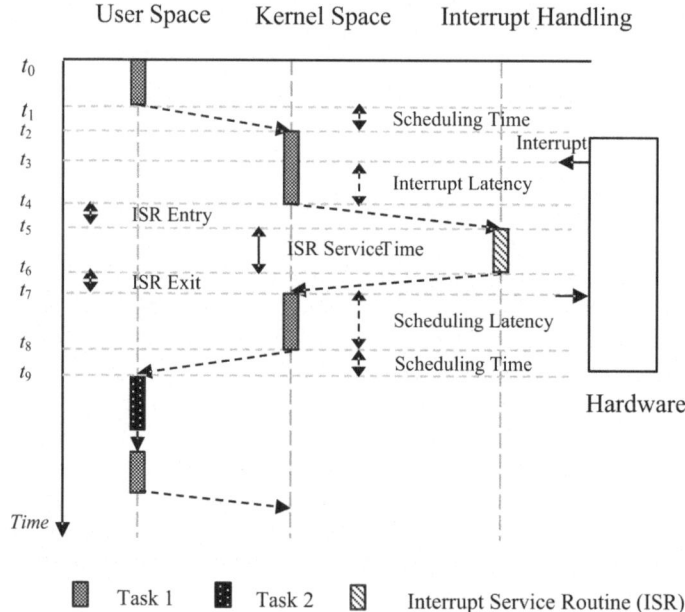

Fig. 1. Task response time and latency in general purpose Linux system

3.3   RTAI Based Ranging System

Taking into account the latency and jitter described above, we select RTAI as the operating system for our TOA ranging system. RTAI is a high performance real-time extension to general-purpose Linux. It achieves real-time task executions by implementing a real-time hardware abstraction layer (RTHAL) on which the real-time application interface is mounted. With this sub-layer, RTAI takes over the system hardware management completely.

Once the RTAI modules are loaded, all hardware interrupts will be intercepted and dispatched by the RTAI and the scheduler in the general-purpose Linux core is also taken over by RTAI's real-time scheduler, which provides simultaneous one-shot and periodic scheduling. RTAI can provide scheduling with a much higher precision than general-purpose Linux because it can schedule tasks based on time stamp counter (TSC) readings. Thus, the theoretical ranging resolution in meters a RTAI sytem can achieve is:

$$d_r = \frac{c}{TSC_{freq}} \tag{2}$$

where d_r is the theoretical minimum distance a RTAI system can measure although the ranging resolution achieved in practice depends on multiple factors, such as channel conditions and hardware response time, c is the speed of light and TSC_{freq} is the TSC frequency which is 1.2G Hz in our ranging system. Besides task scheduling with timers of finer resolutions, RTAI has more advantages as a ranging system over general purpose operating systems. Generally, a RTAI

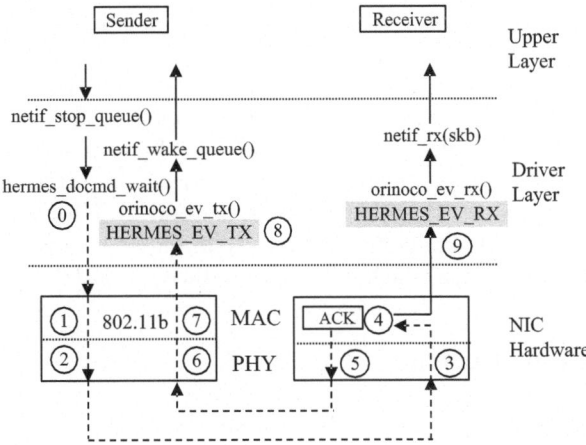

Fig. 2. Tx and Rx path in the WLAN card and driver layer

application is a single-threaded process with a fixed priority. The RTAI thread can be assigned a high priority and run to completion without being preempted. Thus, in comparison with general-purpose operating systems, it can greatly reduce or eliminate the time $t_4 - t_3$ and $t_8 - t_7$ in Fig. 1. Implementation details of RTAI can be found in [9].

Taking into account the above discussion on latency and jitter, the stamping points for RTT in our ranging system are shown in Fig. 2. For easy explanation, we use the same function names as the driver code of our RTT measurement system, which uses an IEEE 802.11b Orinoco Gold card with a Hermes chipset. Drivers for other cards may use different function names, but the same methodology can be used.

Fig. 2 shows the shortest round trip path of a data packet in a TOA ranging system. At the sender, when the driver transmits a packet, it tells the upper layer to stop feeding packet by calling *netif_stop_queue()*. Then, it calls *hermes_docmd_wait()* to transfer data to the MAC layer. In IEEE 802.11 systems, a data frame from the sender requires an ACK frame to be returned to the sender. When the packet is received at the receiver, an ACK frame is assembled and sent to the sender after one SIFS. This ACK is generated by the MAC firmware and is not passed to the upper layer. The MAC layer at the sender raises the *HERMES_EV_TX* interrupt only after it receives this ACK. Thus, the shortest data path for RTT measurement is 0 through 8 (the path with dashed lines in Fig. 2). The receive interrupt, *HERMES_EV_RX* (9 in Fig. 2), is generated only after the data packet is copied to the socket buffer.

To be precise, RTT in our RTAI-based ranging system starts at the time when *hermes_docmd_wait()* is completed; it ends at the time when the interrupt *HERMES_EV_TX* is raised. Hence, Equation 1 can be rewritten as:

$$RTT = t_{tx\_data\_trans} + 2 \times TOA + t_{rx\_ACK\_proc}$$
$$+ SIFS + t_{rx\_ACK\_trans} + t_{tx\_ACK\_proc} \tag{3}$$

where $t_{tx\_data\_trans}$ is the time to transmit a data frame including the preamble, frame header, data payload, and frame extension. $t_{tx\_ACK\_proc}$ and $t_{rx\_ACK\_proc}$ are the time for processing the ACK at the sender and receiver, respectively. $t_{rx\_ACK\_trans}$ is the time to transmit the ACK frame by the receiver.

Equation 3 not only gives the lower bound of RTT, but also identifies where jitters are introduced into ranging results. Particularly, jitters come from the interrupt handling $(0, 8)$, the MAC logic execution $(1, 7, 4)$, and transmitting and receiving $(2, 3, 5, 6)$. These jitters will be eliminated or alleviated by specific data processing algorithms which will be introduced in latter sections.

4   Experiment Setting

A prototype system implementing the ranging logic above is developed on two Dell laptops with a $1.2GHz$ Pentium III mobile processor with no CPU scaling capability. The communication between the sender and receiver is set in ad hoc mode in order to control the jitters at the receiver. The WLAN card used is the Orinoco Gold PCMCIA card which has been set to transmit with the maximum data rate of $11Mbps$. The software used is a vanilla Linux 2.6.23 kernel patched with RTAI 3.6. The test program is implemented in kernel space as kernel modules. RTT data is collected as CPU ticks from the kernel printing buffer. In the data analysis phase, the CPU ticks are converted to nanoseconds according to the following equation:

$$t_{ns} = \frac{Ticks}{TSC_{freq}} = \frac{Ticks}{1.2} \tag{4}$$

where $Ticks$ is time stamp value read from the TSC and TSC_{freq} is the frequency of the TSC in gigahertz.

To verify the performance of the ranging system, tests are conducted both indoor and outdoor. The indoor tests are conducted in a straight aisle inside a building about $3m$ wide. In the outdoor tests, the shortest distance from the test devices to the surrounding buildings is about $10m$. In both situations, the system is placed about 1.5 meters above the ground to preserve the Fresnel zone. Test is repeated with a LOS distance of $0, 15, 30, 45, 60, 75$ feet between the sender and receiver, respectively. In each test, 1000 RTT samples were collected. The RTT collected at distance $0ft$, denoted as RTT_0 is considered as the processing overhead in the ranging system, which corresponds to the summation of the right hand side terms, except for the term $2 \times TOA$, in Equation 3. The distance between the sender and receiver is then computed as:

$$d_i = c \times TOA = c \times \frac{RTT_{d_i} - RTT_0}{2} \tag{5}$$

where RTT_{d_i} is the RTT obtained with distance d_i.

5 Result Analysis

5.1 Data Filtering

Even with time stamping being implemented close to the physical layer, jitters could still be found in the RTT measurement results due to various reasons such as hardware noise and multipath affects. However, with the RTAI ranging system, the data samples, after filtering, are concentrated into a narrow time band of about 10000*ns* for both indoor and outdoor situations.

To remove data outliers, the data is filtered in two steps. The first step is to find the distribution of the data samples. Samples collected are distributed into a series of buckets within a fixed interval (from 95000 ticks to 125000 ticks); any sample falling outside this interval are discarded. The filtered RTT data is shown in Fig. 3.

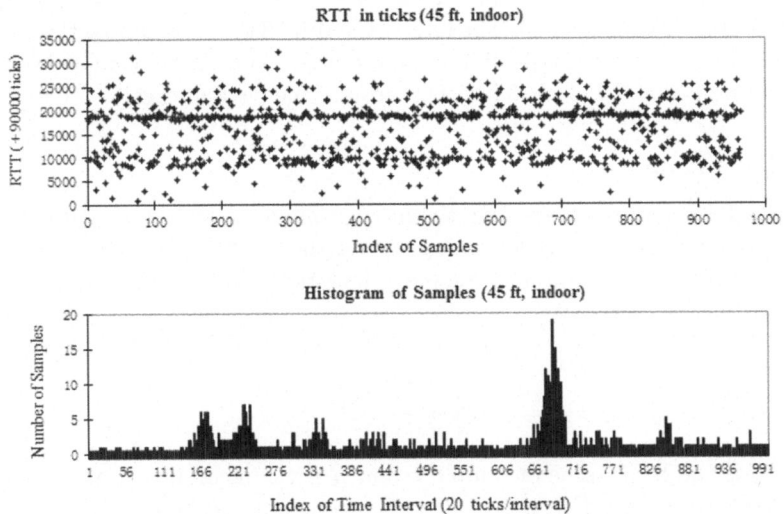

Fig. 3. RTT in ticks at 45*ft* indoor test

Two characteristics can be observed from Fig. 3. First, RTT samples generally follow a *Normal* or a *Gaussian* distribution around the peak value. Thus, statistical data filtering algorithms for *Normal* distribution data can be used for further analysis. Second, several sample clusters exist. We conjecture that the first small clusters on the left are caused by multipath effect.

The second step in data filtering uses a smaller filter window to find the data samples for the final RTT calculation, which are shown as the samples falling within the intervals surrounding the peak interval 683 in Fig. 3. The size of the filter window is experimentally determined. Windows of different sizes are slid around the peak interval and the RTTs that are calculated based on samples within this

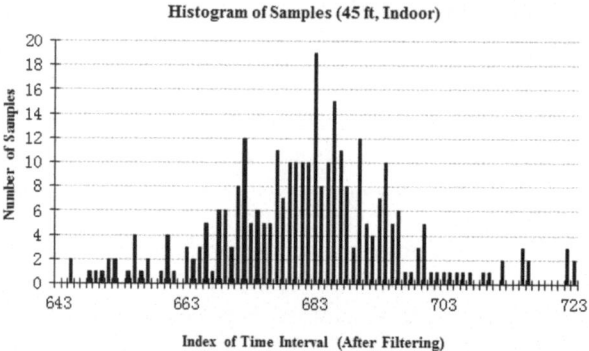

Fig. 4. RTT intervals filtered by sliding window

window are compared. The window size resulting in the smallest error of RTTs are used. For our experiments, a window size of 81 is finally selected (40 intervals to each side of the peak interval). The intervals in Fig. 3 that are filtered this way are shown in Fig. 4, where the peak interval index is 683 and the window boundary intervals are 643 and 723 (±40 intervals around the peak interval).

5.2   Data Analysis

After the RTT sample intervals in the histogram are identified, statistical methods are applied to the samples falling within these intervals. RTT is computed as follows: First, the mean and standard deviation σ are computed for samples falling within the window:

$$\sigma = \sqrt{\frac{1}{N} \sum_{i=1}^{N} (x_i - \mu)^2} \qquad (6)$$

where x_i is the RTT value of sample i, and μ is the mean of all samples within the selected intervals. Estimators of RTT are computed as $\mu \pm \sigma$, $\mu \pm 2\sigma$ and $\mu \pm 3\sigma$, respectively.

Our Experiments show that simply using the mean value as the estimator of RTT has slightly smaller errors. The table in Fig. 5 shows the results of both indoor and outdoor tests using the mean of RTTs:

In the table, RTT_m is the RTT measured in the test, RTT_i is the estimated RTT value for distance i, which is the estimation of the term $2 \times TOA$ in Equation 3. TOA(ns) is computed by Equation 4, and errors are computed as:

$$Err = c\mu/2 - d \qquad (7)$$

where d is the actual distance being measured and c is the speed of light. The table in Fig. 5 shows our RTAI ranging system achieves a result with an error less than $15.19ft$ ($4.63m$) for indoor ranging and $26.1ft$ ($7.95m$) for outdoor ranging. The ranging results compared with the true distances are plotted in Fig. 6.

Indoor (RTT$_0$ = 108175.51 ticks)

Distance(ft)	RTT$_m$ (ticks)	RTT$_i$ (ticks)	TOA(ns)	Est. Dist.(ft)	Error (ft)	Error(%)
15	108244.93	69.42	28.92	28.47	13.47	89.80
30	108254.88	79.37	33.07	32.55	2.55	8.50
45	108302.12	126.61	52.62	51.79	6.79	21.75
60	108358.85	183.34	76.39	75.19	15.19	25.32
75	108366.68	191.17	79.65	78.4	3.4	4.53

Indoor (RTT$_0$ = 106748.23 ticks)

Distance(ft)	RTT$_m$ (ticks)	RTT$_i$ (ticks)	TOA(ns)	Est. Dist.(ft)	Error (ft)	Error(%)
15	106721.17	-27.06	-11.27	-11.10	-26.10	-173.98
30	106803.09	54.86	22.86	22.50	-7.50	-25.01
45	106816.11	67.88	28.28	27.84	-17.16	-38.14
60	106934.67	186.44	77.68	76.46	16.46	27.43
75	106886.14	137.91	57.46	56.56	-18.44	-24.59

Fig. 5. Estimated distance for indoor and outdoor experiments

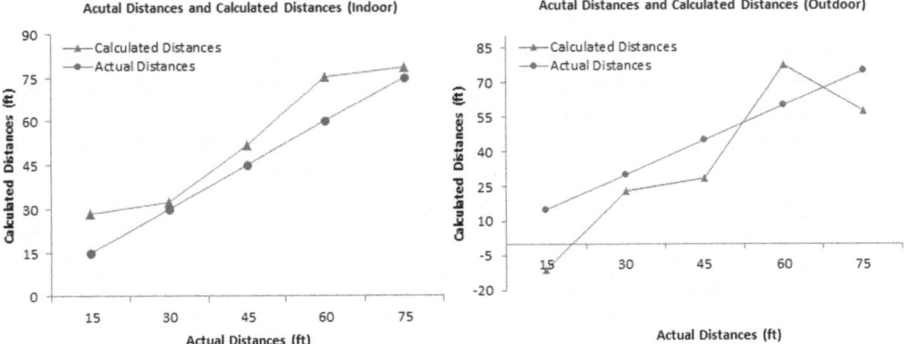

Fig. 6. Distances measured and actual distances for indoor and outdoor experiments

6 Discussion

In our test, RTAI is implemented at both the sender and receiver while the communication mode is set in ad hoc mode. In this configuration, data processing time at the receiver can be well controlled. However, in real world applications, a sender will more likely communicate with an AP. The response of AP depends on many factors that the sender cannot control such as workload change. This may introduce more errors to the ranging result than our configuration. Besides, our experiments are done in a limited number of scenarios, to make this system viable in general application environments, tests in more rigorous situations are needed, such as under different workloads both at the sender and receiver, or under severe channel interference situations. We will conduct these tests in our future work.

7   Conclusion

In this paper, a novel TOA ranging approach using real-time system RTAI is presented. The implemented system is software-based without any special hardware support or modifications to system firmware. RTAI provides real-time guarantees for task executions and has faster response to hardware interrupt than general purpose Linux system. Our experimental results show, with the RTAI real-time system, errors less than $(4.63m)$ for indoor ranging and errors less than $(7.95m)$ for outdoor ranging can be achieved. Our work demonstrated that by exploiting data packet transmission and receive at the instruction execution level at bottom layers of the system, a software based system can achieve good ranging accuracy. Our future work will test the system under more rigorous application conditions.

References

1. Dozio, L., Mantegazza, P.: Real time distributed control systems using RTAI. In: Sixth IEEE International Symposium on Object-Oriented Real-Time Distributed Computing, pp. 11–18 (May 2003)
2. Koyuncu, H., Yang, S.H.: A survey of indoor positioning and object locating system. International Journal of Computer Science and Networks Security 10(5), 121–128 (2010)
3. Liu, H., Darabi, H., Banerjee, P., Liu, J.: Survey of wireless indoor positioning techniques and systems. IEEE Transactions on Systems, Man, and Cybernetics, Part C: Applications and Reviews 37(6), 1067–1080 (2007)
4. Bill, R., Cap, C., Kofahl, M., Mundt, T.: Indoor and outdoor positioning in mobile environments – a review and some investigation on WLAN positioning. Geographic Information Sciences 10(2), 91–98 (2004)
5. Izquierdo, F., Ciurana, M., Barcelo, F., Paradells, J., Zola, E.: Performance evaluation of a TOA-based trilateration method to locate terminals in WLAN. In: 1st International Symposium on Wireless Pervasive Computing, pp. 1–6 (January 2006)
6. Karalar, T., Rabaey, J.: An RF tof based ranging implementation for sensor networks. In: IEEE International Conference on Communications, vol. 7, pp. 3347–3352 (June 2006)
7. Golden, S., Bateman, S.: Sensor measurements for Wi-Fi location with emphasis on time-of-arrival ranging. IEEE Transactions on Mobile Computing 6(10), 1185–1198 (2007)
8. Ciurana, M., López, D., Barceló-Arroyo, F.: SofTOA: Software Ranging for TOA-Based Positioning of WLAN Terminals. In: Choudhury, T., Quigley, A., Strang, T., Suginuma, K. (eds.) LoCA 2009. LNCS, vol. 5561, pp. 207–221. Springer, Heidelberg (2009)
9. http://www.aero.polimi.it/~rtai
10. Muthukrishnan, K., Koprinkov, G., Meratina, N., Lijding, M.: Using time-of-flight for WLAN localization: feasibility study. Center for Telematics and Information Technology (WLAN) technical report, TR-CTIT-06-28 (June 2006)

Power Reduction in WDM Mesh Networks Using Grooming Strategies

Farid Farahmand[1], M. Masud Hasan[2], and Joel J.P.C. Rodrigues[3]

[1] Sonoma State University, Rohnert Park CA 94928, USA
farid.farahmand@sonoma.edu
http://www.sonoma.edu/users/f/farahman/
[2] Dept. of Math. & CS, Elizabeth City State University, North Carolina, USA
mmhasan@mail.ecsu.edu,
[3] Instituto de Telecomunicacoes, University of Beira Interior, Covilha, Portugal
joeljr@ieee.org

Abstract. This work reports on the benefits of using energy-efficient grooming strategies in WDM mesh networks in terms of the overall network power consumption. We examine a key enabling node architecture called tap-or-pass (TOP) and demonstrate how it can support lightpath *extension* and lightpath *dropping*. Using these grooming concepts we propose several grooming strategies. Through extensive simulation, we demonstrate that, given a network with dynamic traffic requests, the proposed grooming strategies lead to considerable energy saving and comparable request blocking, in particular when the network load is moderate.

Keywords: Green Networking, Energy Efficiency, Optical Networks, Routing and Traffic Grooming.

1 Introduction

As the Information and Communication Technology (ICT) infrastructure grows more electrical power will be required to support telecommunication networks. In fact, it is estimated that in the developed countries approximately five percent of the total electrical energy is consumed by the telecommunication and IT industry [1]. In the developing nations, ICT infrastructure consumes approximately one percent of the total electricity consumption [2] and as higher-speed national broadband access networks grow, this number is expected to increase exponentially. Consequently, a large body of works have concentrated on studying and reducing energy consumption in critical areas such as chip design [3,4], wired-line networks [5,6], and servers and application [7]. However, only limited attention has been devoted to study energy efficient optical networks. In fact, while optical networks continue to be the champion of future networks (e.g., 400 GigE and 1 TbE [8]) due to their high capacity, low transmission loss, transparency to signal rate and format, and resilience to noise and environmental harsh conditions, their power consumption is substantially high and manifests considerable operational cost.

Joel J.P.C. Rodrigues et al.: (Eds.): GreeNets 2011, LNICST 51, pp. 99–113, 2012.
© Institute for Computer Sciences, Social Informatics and Telecommunications Engineering 2012

In recent years, a number of studies, such as [9, 10] and [11], have focused on comparing the the power consumption of photonic and electrical subsystems in an optical cross-connect in WDM networks. In their survey, the authors of [12] provide a detailed overview of energy conservation approaches across core, metro, and access levels of optical networks.

In study of energy-efficient optical networks, a number of efforts have been dedicated to develop energy-aware grooming mechanisms, including [13–16]. One potential issue with the above traffic grooming proposals is that they require activating power-hungry electronic multiplexers and demultiplexers in the electronic layer of optical cross-connects. Thus, in order to support traffic grooming, the dropped optical traffic, or *lightpath*, has to be converted into electrical form using optical-electrical (OE) converters. Furthermore, the remaining traffic must be converted back to optical form using electrical-optical (EO) converters prior to retransmission of the lightpath toward downstream nodes. Such signal conversion and switching is known to be power intensive [17].

In this work we focus on energy-aware traffic grooming in WDM networks where lightpaths are shared between multiple requests, and local traffic is only dropped optically. Thus, the dropped traffic at intermediate nodes no longer has to go through electronic devices in order to be added and retransmitted. The key enabling technology to support optical dropping at intermediate nodes, while passing the lightpath to other nodes, is the tap-or-pass (TOP) node architecture. In this architecture, the incoming lightpaths can pass through intermediate nodes *or* have their power split unequally so that a small portion of the optical energy is dropped while the rest of the energy is passed to the subsequent downstream nodes. Similar node architectures have been discussed for various applications in previous literature, including [18].

Motivated by our earlier work in [19], [20], and [21], we demonstrate the benefits of exploiting the TOP architecture to support energy-aware traffic grooming in optical networks. Using TOP-enabled nodes, we evaluate the energy saving benefits of two grooming concepts called *lightpath dropping* and *lightpath extension*. We refer to these as *lightpath-based grooming* (LBG). In the lightpath dropping approach, a lightpath can pass through an intermediate node, while partially being dropped. Lightpath extension on the other hand, is based on optically extending an existing lightpath beyond its original terminating node. The key contribution of this work compared to [20] is that we present the details of the auxiliary graph and how the LBG algorithms are implemented when traffic requests arrive dynamically.

Through simulations, we evaluate the performance of lightpath dropping and extension using our proposed grooming algorithms in terms of energy usage and request blocking probability. We compare our results with traditional grooming techniques as described in [22].

The rest of this paper is organized as follows. In Section 2, we describe details of the TOP node architecture. In Section 3, we model the TOP-enabled network and mathematically compare lightpath dropping and lightpath extension in terms of power budget. In Section 4, we describe the grooming algorithms and our proposed auxiliary graph. Then, in Section 5, we provide our simulation results. In Section 6, we conclude the paper.

2   Network Architecture and Problem Formulation

In general, a WDM node with grooming capability has a two-layer architecture composed of an optical cross-connect (OXC) and an electronic switch. The OXC may be accompanied with wavelength converters, in which the incoming wavelengths can change color prior to leaving the switch.

A typical electronic switch layer in a WDM node with grooming capacity consists of electronic-optical-electronic (EOE) and optical-electronic-optical (OEO) converters, multiplexers and demultiplexers, and a grooming switch fabric. Each multiplexer, in turn, may have one or more transmitters and receivers, each connected to an add or drop port on the OXC, respectively. Add/drop ports, shown in Fig. 1(a), allow the lower rate signals to be inserted (or extracted) into the high-speed optical signals.

2.1   TOP Architecture

In this paper we use a modified OXC, as shown in Fig. 1(b). In this architecture an incoming lightpath can pass through the OXC (maintaining its full power) *or* tap-or-pass using a passive unbalanced splitter. Using the tap-or-pass (TOP) architecture, only a small portion of the incoming optical optical power is dropped and the rest can be sent to the next node. When a lightpath is terminated at a node, it is simply tapped and the rest of the energy can be ignored. As noted before, when wavelength conversion capability is available, the continuing portion of the optical signal can be carried using a different color.

One drawback of the TOP architecture is that it reduces link utilization by carrying extra traffic. This is because the aggregated traffic will be reaching all the nodes visited by the lightpath. Another issue with the TOP architecture is that as lightpaths are tapped their optical power is reduced, and thus, they will have shorter reach. The main motivation in implementing the TOP architecture, however, is that it reduces the overall energy consumption of the node because it no longer requires turning on the electronic multiplexers (including transmitters and OEOs) of a node in order to *retransmit* the remaining dropped traffic.

The TOP architecture offers two distinct features: *lightpath dropping*, which refers to the ability to drop a lightpath on the intermediate nodes, and *lightpath extension*, which refers to extending a lightpath beyond its current terminating node to a new node. These features are depicted in Fig. 2. A typical point-to-point lightpath between a source-destination node pair is shown in Fig. 2(a), where the lightpath is terminated at Node 1. Fig. 2(b) shows how the same lightpath can be dropped at intermediate Node 0. Lightpath extension to Node 2, beyond the original terminating node (Node 1) is demonstrated in Fig. 2(c). A combination of lightpath dropping and lightpath extension is shown in Fig. 2(d). An important feature of the TOP architecture is that no traffic interruption

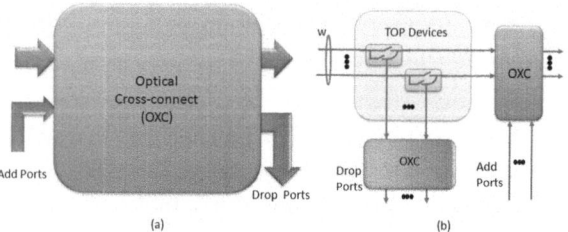

Fig. 1. Optical crossconnect of a WDM node with (a) add/drop ports and (b) with tap-or-pass (TOP) capability

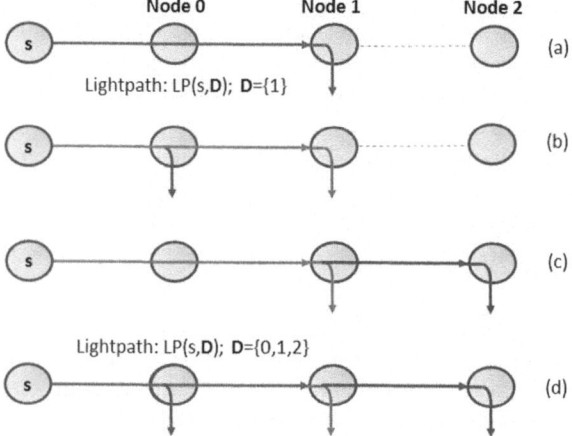

Fig. 2. A network with (a) point-to-point lightpath connections; (b) lightpath dropping at an intermediate node; (c) lightpath extension; and (d) combination of lightpath dropping and extension

occurs as a lightpath is dropped or extended to other nodes, as long as all nodes are receiving appropriate amount of optical power.

2.2   Problem Formulation

The energy-efficient traffic grooming problem formulation can be stated as follows. Given the network topology with all nodes being TOP-enabled, the number of wavelengths per link, the number of transceivers at each node, and the incoming traffic request, find the routing and wavelength assignment for the requested traffic demand on the virtual topology such that minimum power consumption is achieved in the network. In the following sections we demonstrate how lightpath-based grooming (LBG) can be utilized to improve the overall energy saving of the network without degrading its performance.

3   Power Budget Model

A critical aspect of implementing lightpath-based grooming is ensuring a minimum power level at all nodes. Thus, as a lightpath is extended or dropped at other nodes, the power budget must be recalculated to ensure that all nodes receive a sufficient amount of optical power. In this section, we provide a power budget model for LBG. Using this model it is possible to evaluate the feasibility of LBG.

We assume that a lightpath between a source-destination node pair (s, d), denoted by $LP(s, d)$, can be dropped at any node k, including an intermediate node, the destination node d, or an extended node one or more hops away from node d. Therefore, $LP(s, D)$, can be modeled as shown in Fig. 3, starting at node s and passing through a set of nodes, $D = [0, 1, 2, ..., k]$. We define P_k as the amount of power available towards the downstream node and $P_k \cdot G_k^{-1} \cdot (\frac{\alpha}{1-\alpha})$ as the amount of power received by any intermediate or ending node k along a given lightpath. In this expression, α is the portion of power dropped at each node's receiver ($0 < \alpha < 1$), and G_k is the optical power amplification performed by node k prior to passing the lightpath to the next node. In general, the power budget model using TOP-enabled switches can be expressed as

$$P_k = [(P_{k-1} - L_k) \cdot (1 - \alpha.s_k)] \times G_k \ \forall k \in D, k \neq s. \tag{1}$$

In the above formulation, L_k is the power lost in the link connecting two consecutive nodes $k - 1$ and k, due to attenuation. We define matrix $S = [s_0, s_1, ..., s_k]$ as an indicator in which $s_k = 1$ represents the drop of lightpath at node k and $s_k = 0$ indicates that the lightpath is passing through node k and no splitting is performed. For example, in Fig. 2(c), $S = [0, 1, 1]$ indicates that the lightpath is being dropped at Node 1 and Node 2. In our analysis, we assume that all nodes have the same gain, G, and link loss, L. We denote the initial output optical power of a lightpath at a source node by P_s.

The amount of power *dropped* at each node k must be at least P_{min}, depending on the sensitivity of the node's receiver:

$$\frac{P_k}{G_k} \cdot (\frac{\alpha}{1 - \alpha}) \geq P_{min}. \tag{2}$$

Therefore, as a lightpath is dropped on an intermediate node, the power received by the last node on the lightpath does not fall below P_{min}. On the other hand, an existing lightpath can only be *extended* by h hops from node d to $\dot{d} = d + h$, without any type of amplification, $G = 1$, if

$$\frac{P_{\dot{d}}}{G_{\dot{d}}} . (\frac{\alpha}{1 - \alpha}) = P_d \cdot (h \cdot L) \geq P_{min}. \tag{3}$$

For example, in Fig. 2(c) $h = 1$ and $P_{\dot{d}}$ refers to the available power at Node 2.

Referring to Eqn. 1, it is possible to rearrange the expression as follows:

$$P_k = [(1 - \beta_k) \cdot (1 - \alpha.s_k)] \times G_k \times P_{K-1} \ \forall k \in D, k \neq s, \tag{4}$$

where we define $\beta_k = L_k / P_{k-1}$ as the *link loss ratio*. Clearly, $0 < \beta_k < 1$. A closer look at the above expression suggests that, for a given lightpath between two or more nodes, when the dropped portion of power at each node is much larger compared to the link loss ratio, $\beta_k >> \alpha$, lightpath dropping is more power efficient compared to lightpath extension. That is, for an existing lightpath $LP(s, d)$, lightpath dropping results in larger available power at the terminating node d to be forwarded downstream.

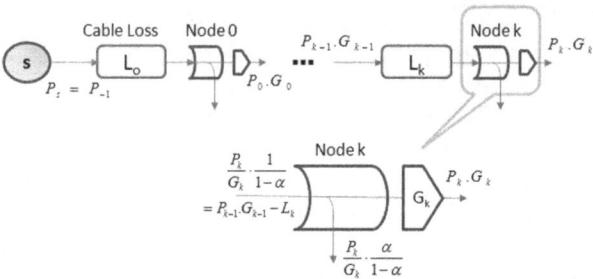

Fig. 3. Power budget model for supporting lightpath extension and lightpath dropping using TOP-enabled node architecture

In our analysis, we assume that the output power is such that $P_s \geq (h_{max}.L) \cdot \alpha$, where h_{max} is maximum hop distance of the network. In other words, all lightpath has sufficient power to be terminated at any node in the network.

4   Lightpath-Based Grooming (LBG)

In this section we discuss the implementation of a lightpath-based grooming, including lightpath dropping and lightpath extension. The principle of the LBG is similar to the auxiliary graph described in [19]. However, the key difference is that we combine the graph model with the power budget model, described in Section 3.

4.1   Auxiliary Graph Model

Given a network with N nodes and W wavelengths per fiber link, the physical network can be represented by a graph $G_p = (V_p, E_p)$. In this representation, V_p is the set of network nodes, and E_p is the set of links connecting the nodes. The current status of the network can be modeled by a W-layer auxiliary grooming graph, $GG = (V, E)$, where each layer corresponds to the state of a wavelength in the network. A vertex $v \in V$ in the auxiliary graph, GG, represents the optical receiving or transmitting capabilities of a physical node on a particular wavelength layer. Therefore, a physical node can be represented by W receiving and W transmitting vertices.

On the other hand, E is a set of weighted directional edges which corresponds to available optical paths between node pairs. In our graph model, we define two basic edge types, namely, *grooming* edges and *optical* edges. A grooming edge abstracts the node's grooming capacity enabling an optical signal to be dropped and processed electronically. Therefore, for each physical node, there will be one grooming edge between a single receiving vertex and each transmitting vertex. We denote a grooming edge from a receiving vertex on layer x to a transmitting vertex on layer y on node i by $GP_i^{x,y}$.

An optical edge, on the other hand, represents an all-optical path between a node pair. Depending on the node architecture, in our graph model, we define the following optical edges, which can be established between a node pair (i, k) with one intermediate node j or more, on wavelength layer w:

- *Existing lightpath*, LP_{ik}^w, describing an active lightpath currently carrying traffic between nodes i and k;
- *Potential lightpath*, PLP_{ik}^w, representing one or more available wavelength links, which can support a new lightpath from node i to k;
- *Potential extended lightpath*, $PELP_{ik}^w$, expressing an existing lightpath, LP_{ij}^w, which can potentially traverse optically beyond its current end node, j, and reach node k through one or more available wavelength links;
- *Sub-lightpath*, SLP_{ij}^w, describing a possible optical connection between the source node, i, and an intermediate node, j, of the existing lightpath, LP_{ik}^w.

Note that for each existing lightpath with I intermediate nodes, there will be as many as I sub-lightpaths, all having the same free capacity. These concepts are illustrated in Fig. 4(a).

The lightpath-based grooming algorithm with intermediate dropping and extension capacity (LPwDwE) supports two basic operations in order to route new connection requests: (1) existing lightpaths can be dropped at their intermediate nodes, while continuing their path to the end node; (2) existing lightpaths can be extended beyond their end nodes. These concepts are shown in Fig. 4(b). The main motivation for implementing the LPwDwE is to provide higher flexibility in finding the most appropriate routing path between a node pair.

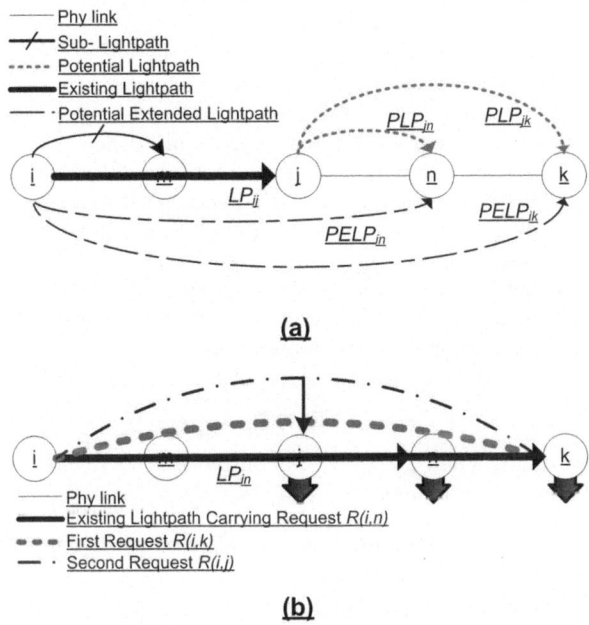

Fig. 4. (a) Illustration of different optical edges used in the auxiliary graph; (b) The LPwDwE algorithm allows lightpath extension to node k and dropping on intermediate nodes j and n

The LPwDwE algorithm consists of two basic routines: *ReqSetup* and *ReqTeardown*. For each new connection request, the *ReqSetup* routine constructs a new auxiliary graph representing the current status of the network and finds the shortest path between the requested node pair. Details of the *ReqSetup* routine upon arrival of a new request $Req(s, d, B)$, where s and d are the source and destination nodes, respectively, and B is the request's demand, are described in Table 1.

Table 1. Algorithm description for the *ReqSetup* routine in LPwDwE

For a given request $Req(s, d, B)$:

1. For each wavelength layer w and each node i on the physical graph G_p
   (a) Find the shortest path between node i and every other node j, such that a potentially new lightpath can be established between the two nodes, PLP_{ij}.
   (b) For every existing lightpath, between nodes i and j, LP_{ij}, with free capacity $C_f \geq B$,
      i. Find all possible sub-lightpaths between node i and all the intermediate nodes on LP_{ij}.
      ii. Find all possible potential lightpaths by extending LP_{ij} on available links.
   (c) Assign weight to all edges including potential lightpaths, potential extended lightpaths, existing lightpaths, sub-lightpaths, and grooming edges according to the grooming policy.
2. Search for the shortest path on the auxiliary graph between node s and d. If no such path was found, discard the request; otherwise, continue to next step.
3. Set up the route for the request $Req(s, d, B)$ and update the network status to reflect the latest connections and available resources.

On the other hand, when a request is completed the *ReqTeardown* routine is executed and operates as follow:

− *Step 1*: The request's demand is removed from all lightpaths carrying the request;
− *Step 2*: All *inactive* wavelength links along lightpaths carrying the request are removed. If all wavelength links on a lightpath are inactive, the entire lightpath will be removed;
− *Step 3*: The network state is updated accordingly to represent the latest available resources.

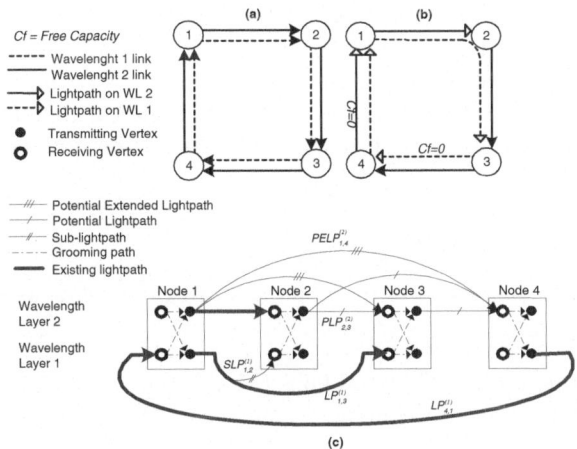

Fig. 5. (a) An example of a four-node network with two wavelengths in each fiber; (b) the current state of the network with available wavelength links; (c) the auxiliary graph, GG, for the current state of the network shown in (b).

4.2 Example

We illustrate the above concepts by means of an example. Fig. 5(a) shows a four-node network with four unidirectional fiber-links, each having two wavelengths. Each node is equipped with two transmitters and two receivers and has full-grooming capacity (the entire incoming data can be groomed). Initially, we assume that no connections exist in the network. Fig. 5(b) shows the current state of the network after a number of connection requests are established. Upon arrival of a new request, the auxiliary graph, shown in Fig. 5(c) can be established. We assume $LP_{3,4}^1$ and $LP_{4,1}^2$ have no available bandwidth and thus they are not shown in the auxiliary graph. Using the two available wavelength links between node pairs (2,3) and (3,4), we can generate 3 distinct potential lightpaths on Layer 2. The existing lightpath between node pair (1,2) can also be extended to Nodes 3 and 4. Furthermore, the existing lightpath on wavelength Layer 1 between Nodes 1 and 3 can support a sub-lightpath between node pairs (1,2), denoted by $SLP_{1,2}^1$. Let us assume that Node 3 requests a new connection to Node 2. Based on available resources, indicated by the auxiliary graph in Fig. 5(c), this request can be satisfied through the following shortest multi-hop path: $PLP_{3,4}^2$, $GP_4^{2,1}$, $LP_{4,1}^1$, $GP_1^{1,2}$, and $LP_{1,2}^2$.

4.3 Algorithm Complexity

The complexity of LPwDwE is mainly attributed to the *ReqSetup* routine, which in turn is directly tied to complexity of the shortest path algorithm. For example, assuming we implement Dijkstra's shortest path algorithm, the worst-case complexity of the *ReqSetup* will be equivalent to finding all available shortest paths between all nodes on all wavelength layers *and* the shortest path for the $Req(s, d, B)$ among all layers between the node pair (s, d). Thus, the worst-case complexity will be equivalent to $O(wn^3) + O((nw)^2)$. Note that if the number of wavelengths is much larger than the number of nodes in the network, as is the case in backbone networks with dense WDM links, the dominating factor will be $O((nw)^2)$.

4.4   Grooming Strategies

In our study, we consider several grooming strategies. Below we briefly describe each one.

- Minimize the number of logical hops (MinLH), i.e., minimize electronic processing for connection requests. In this case, the total cost to establish a connection will be based on the number of logical hops.
- Minimize the number of physical hops (MinPH), i.e., maximize the wavelength utilization. Thus, the total end-to-end cost will be equivalent to the number of physical hops between the source-destination node pair.
- Minimize the number of new lightpaths (MinNL), i.e., minimize the number of transmitters and receivers.
- Minimize the number of physical hops on lightpaths carrying the request (MinTH), i.e., maximize the wavelength utilization. In this case, the weight assignment for all optical links is equivalent to the number of physical hops on the entire edge, including the ones beyond the destination node.

Using the auxiliary graph, when several shortest paths are available for a single connection request, a secondary objective is chosen to select the most appropriate available route. For example, assuming that the main objective is to minimize the number of logical hops, and more than one such route is available, the route with the least number of logical hops is selected.

When limited resources are available, the above LBG strategies lead to a different utilization of the network resources and, thus, to a different level of network performance, in terms of request blocking probability and energy consumption.

5   Performance Analysis

In this section, we describe simulation results obtained by implementing the aforementioned algorithms in the TOP architecture. The schemes are evaluated in terms of blocking and energy consumption. In our simulation, we consider the 14-node NSF network with 21 bidirectional links. We assume each link supports 4 wavelengths in each direction, operating at OC-192 rate. Connection requests are generated dynamically, following a Poisson process with uniform distribution among the node pairs. The connection requests are uniformly distributed among OC-3, OC-12, or OC-48 rates. We assume that there is no wavelength conversion.

Unless otherwise mentioned, there are 4 transmitters and 4 receivers in each node. The power consumption of the grooming module (transmitter or receiver, including the E/O and O/E) is 160 W/module [23]. This figure is based on the power consumption figure of the Cisco Catalyst 6500 series published in [24]. For our analysis we assume that only 5% of the power is dropped at each node and that the remaining of power will pass through the TOP device. We assume that all unused modules (i.e., modules not processing traffic) are in idle mode and consume a negligible amount of power. In order to obtain the following results, we took the average of 105 experiments for each data point, presenting a 95% confidence interval.

5.1   Comparing Grooming Strategies

We first examine the performance of LBG strategies. Then, we compare the performance of lightpath extension and lightpath dropping for a given grooming strategy.

Fig. 6 displays the blocking probability obtained by implementing different grooming strategies, namely, MinLH, MinPH, MinNL, MinTH, as a function of the load, when the overall network load varies from 20% to 98% of the network capacity. In this case, we assume each node can perform *both* lightpath extension and lightpath dropping. This figure shows that, regardless of the grooming strategy, when the number of transmitters and receivers are limited, implementing TOP can improve the overall blocking probability when compared to traditional grooming (NoTP), particularly at lower loads. The reason that none of the grooming strategies performs better than NoTP at higher loads is due to the fact that lightpath dropping and lightpath extension carry the entire groomed traffic to all intermediate and extended nodes, thus lowering the network utilization.

Fig. 6 indicates that when the number of receivers and transmitters is limited, the best performance is achieved using the MinTH grooming policy. This figure also suggests that, among the proposed grooming policies, the poorest performance is achieved by MinNL. Note that, using MinNL, the LBG strategy attempts to use available resources before establishing new lightpaths.

Fig. 7 depicts the total energy usage in kilowatt-hour when LBG strategies are implemented, compared to the traditional grooming (NoTP). We note that, based on our results, all the proposed grooming strategies perform consistently better then NoTP in terms of energy efficiency.

Comparing Fig. 6 and Fig. 7, we observe that, in general, for moderate loads, NoTP experiences higher blocking and energy consumption. We also observe that MinNL strategy experiences a high blocking and high energy consumption compared to other grooming strategies. On the other hand, MinLH appears to perform the best in terms of both network performance and energy consumption.

Fig. 8 shows the percentage of annual energy savings in the network with the TOP architecture when compared to traditional grooming. In our calculation of energy savings, the power consumption for cooling (e.g., air conditioning the building) is not considered. Furthermore, it is assumed that the electronic switch fabric is always fully active, independent of the traffic load. We believe these two assumptions make the energy saving results very conservative. Our results indicate that energy savings of up to 20% can be achieved by exploiting TOP

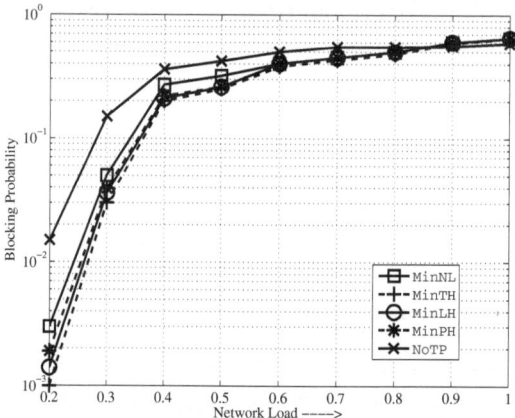

Fig. 6. Blocking probability comparison between different grooming strategies when both lightpath dropping and extension are allowed

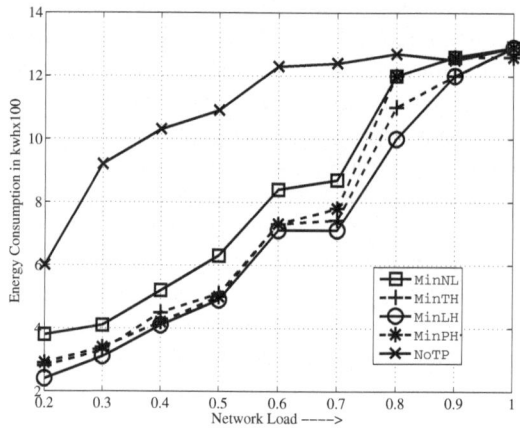

Fig. 7. Energy consumption comparison between different grooming strategies when both lightpath dropping and extension are allowed

architectures when the network load is low. As the network load increases, the impact of using TOP architecture is reduced and eventually all the electronic devices in nodes must be activated.

5.2 Comparing Lightpath Dropping and Extension

In this section we compare the performance of lightpath dropping and lightpath extension in terms of blocking and energy consumption. Fig. 9 compares the blocking probability using MinLH for the following cases: lightpath dropping only (WD-NE); lightpath extension only (ND-WE); and both lightpath extension and light dropping (WD-WE). Our results indicate that, in general, WD-WE consistently performs better over different load values, particularly when the load is low.

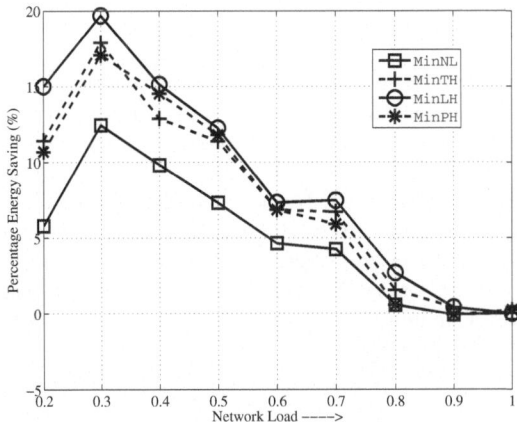

Fig. 8. Annual energy saving in dollars compared to NoTP

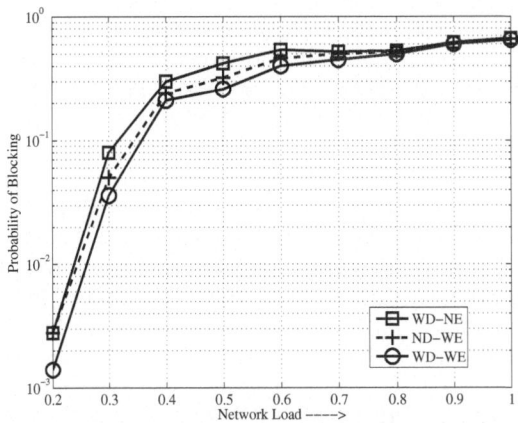

Fig. 9. Comparing blocking probability with MinLH using lightpath dropping only (WD-NE) lightpath extension only (ND-WE), and both lightpath extension and light dropping (WD-WE)

6   Conclusion

In this work we examined the energy saving benefits of tap-or-pass (TOP) node architecture in optical mesh networks. Using this architecture we reported on performance comparison between different lightpath-based grooming strategies. Our results demonstrate that in general, when the number of tributary transmitters and receivers is limited, lightpath-based grooming using the TOP paradigm can perform better in terms of blocking probability, particularly, when the network load is low. Furthermore, the proposed lightpath grooming policies can offer moderate energy savings and thus, operational expenditure reduction.

References

1. Smart 2020: Enabling the Low Carbon Economy in the Information Age. Global eSustainability Initiative (GeSI) (February 2010), http://www.gesi.org/
2. Hinton, K., et al.: Power consumption and energy efficiency in the Internet. IEEE Network 25, 6–12 (2011)
3. Pering, T., Burd, T., Bordersen, R.: The Simulation and Evaluation of Dynamic Voltage Scaling Algorithms. Proceedings of the International Symposium on Low Power Electronics and Design (August 1998)
4. Raghunathan, V., Srivastava, M., Gupta, R.: A Survey of Techniques for Energy Efficient On-Chip Communication. In: Proceedings of Design Automation Conference, Anaheim, CA (June 2003)
5. Gupta, M., Singh, S.: Greening of the Internet. In: Proceedings of ACM SIGCOMM 2003 (August 2003)
6. Gupta, M., Singh, S.: Energy conservation with low power modes in Ethernet LAN environments. In: Proceedings of IEEE INFOCOM (minisymposium) (May 2007)
7. Tornatore, M., Zhang, Y., Chowdhury, P., Mukherjee, B.: Energy Efficiency in Telecom Optical Networks. IEEE Communications Surveys (2010)
8. 40 Gigabit Ethernet and 100 Gigabit Ethernet, http://www.ethernetalliance.org/ (accessed March 2010)
9. Tucker, R.S.: Optical Packet Switching: A Reality Check. Optical Switching and Networking 5(1), 2–9 (2008)
10. Murakami, M., Oda, K.: Power consumption analysis of optical cross-connect equipment for future large capacity optical networks. In: 11th International Conference on Transparent Optical Networks, ICTON 2009, pp. 1–4 (July 2009)
11. Tucker, R.S.: The Role of Optics and Electronics in High-Capacity Routers. IEEE/OSA Journal of Lightwave Tech. 24(12) (December 2006)
12. Zhang, Y., Chowdhury, P., Tornatore, M., Mukherjee, B.: Energy efficiency in telecom optical networks. IEEE Communications Surveys and Tutorials (July 2010)
13. Yetginer, E., Rouskas, G.N.: Power efficient traffic grooming in optical WDM networks. In: Proceedings of IEEE GLOBECOM, pp. 1–6. IEEE (2010)
14. Huang, S., Seshadri, D., Dutta, R.: Traffic grooming: A changing role in green optical networks. In: Proceedings of IEEE GLOBECOM, pp. 1–6. IEEE (2010)
15. Hasan, M.M., Farahmand, F., Patel, A., Jue, J.P.: Traffic Grooming in Green Optical Networks. In: IEEE International Conference on Communications (ICC), South Africa (May 2010)
16. Farahmand, F., Hasan, M.M., Cerutti, I., Jue, J.P., Rodrigues, J.: Differentiated Energy Saving Grooming in Optical Networks. In: Proc. of IEEE GLOBECOM, Miami, Fl (December 2010)
17. Cerutti, I., et al.: Power saving architectures for unidirectional WDM rings. In: IEEE/OSA Conference on Optical Fiber Communication, San Diego, CA (May 2009)
18. Ali, M., Deogun, J.: Power-efficient design of multicast wavelength-routed networks. IEEE Journal on Selected Areas in Communications 18, 1852–1862 (2000)
19. Farahmand, F., et al.: Efficient Online Traffic Grooming Algorithms in WDM Mesh Networks with Drop-and-Continue Node Architecture. In: Proc. of IEEE Broad-Nets, Boston, MA, pp. 160–174 (October 2004)
20. Hasan, M.M., Farahmand, F., Cerutti, I., Jue, J.P.: Lightpath-Based Grooming in WDM Networks. In: IEEE ICCCN, Maui (July/August 2011)

21. Hasan, M.M., Farahmand, F., Cerutti, I., Jue, J.P.: Energy-efficiency of Drop-and-Continue Traffic Grooming. In: IEEE/OSA Optical Fiber Communication Conference, Los Angeles, CA (March 2010)
22. Zhu, K., Mukherjee, B.: On-line approaches for provisioning connections of different bandwidth granularities in WDM mesh networks. In: IEEE/OSA Conference on Optical Fiber Communication, pp. 549–551 (March 2002)
23. Aleksic, S.: Energy Profile Aware Routing. In: IEEE/OSA Conference on Optical Fiber Communication, pp. 245–258 (August 2009)
24. Cisco® Power Calculator, tools.cisco.com/cpc (accessed: February 2009)

Power Consumption Analysis
of Data Center Architectures

Rastin Pries, Michael Jarschel, Daniel Schlosser, Michael Klopf, and Phuoc Tran-Gia

University of Würzburg, Institute of Computer Science,
Chair of Communication Networks, Würzburg, Germany
{pries,michael.jarschel,
schlosser,trangia}@informatik.uni-wuerzburg.de

Abstract. The high power consumption of data centers confronts the providers with major challenges. However, not only the servers and the cooling consume a huge amount of energy, but also the data center network architecture makes an important contribution. In this paper, we introduce different data center architectures and compare them regarding their power consumption. The results show that there are some differences which should not be neglected and that with only minor modifications of the architecture, it is possible to save a huge amount of energy.

Keywords: data center, energy efficiency, networking.

1   Introduction

Data centers are attracting more and more interest, offering a large variety of services such as online gaming, data storage, data processing, and online office products. However, there are still a lot of challenges to be solved, e.g., overall performance, energy efficiency, resilience, scalability, and how to transport the data to the consumer. Most data center providers currently focus on building their data centers only with commercial off-the-shelf (COTS) hardware to reduce the cost and to be easily maintainable. In addition, the data center should be easily extensible and should scale up to 100,000 servers. Therefore, the new data centers are compromised of containers, each carrying up to 2,500 servers.

Besides this information, most cloud providers keep their data center architectures as a secret. Only facebook lately set up the Open Compute Project [1] releasing their open hardware especially designed for data centers. However, the data center network architecture is not yet released and it is stated that they work within the also newly created Open Networking Foundation [2] to create a new, energy efficient data center network architecture.

There are several ways of how to reduce the power consumption in a data center, ranging from energy efficient server hardware as proposed by facebook over coordinated cooling and load management to full virtualization. Generally, the energy efficiency of a data center is measured using the Power Usage Effectiveness (PUE) which was developed by the green grid consortium. The PUE is computed as follows:

$$PUE = \frac{\text{total facility power}}{\text{IT equipment power}}. \tag{1}$$

Joel J.P.C. Rodrigues et al.: (Eds.): GreeNets 2011, LNICST 51, pp. 114–124, 2012.
© Institute for Computer Sciences, Social Informatics and Telecommunications Engineering 2012

An ideal PUE is 1.0, whereas the state-of-the-art industry average is 1.5. The new facebook data center has a PUE of 1.07 calculated at full load over an 8 hour period in December 2010.

In this paper, we instead focus on the power consumption of data center network architectures and evaluate the currently deployed architectures and some proposed architectures according to their power consumption. We evaluate the following six architectures, two-tier, three-tier, DCell, BCube, fat-tree, and elastic-tree. So far, these architectures have only been compared by Wu et al. [3] regarding the number of necessary switches, cables, etc. and Chen et al. [4] provided an overview of routing in data centers, also considering energy-efficiency on the routing layer. Another paper looking at the power consumption of today's data centers is proposed by Poess and Nambiar [5]. In the paper, a power consumption estimation model for TPC-C benchmarks is proposed. The model is applied to published TPC-C benchmarks and the performance and energy performance trends are shown. The only paper looking at a similar direction as in this paper is published by Gyarmati and Trinh [6]. Unfortunately, their power consumption figures only show some isolated results and thus, the architectures are difficult to compare. In contrast to their publication, we evaluate more data center architectures and show the power consumptions for architectures from a few server to up to 70,000 servers.

The remainder of the paper is structured as follows. In Section 2, we describe the evaluated data center architectures. Section 3 shows the used parameters for the evaluation of the data center architectures. The results from the performance evaluation are described in Section 4. We conclude the paper by summarizing our main contributions in Section 5.

2   Data Center Architectures

Several different network architectures have been proposed for data centers ranging from switch-centric approaches such as butterfly, Clos network, and VL2 to server-centric approaches such as mesh, torus, star, ring, hypercube, DCell, and BCube. In this paper, we only focus on the most promising and well-known approaches and evaluate their impact on the total power consumption. All six considered architectures are introduced in the following.

2.1   Two-Tier Architecture

A two-tier data center architecture is shown in Figure 1. The servers are arrange into racks and form together with the Top of Rack (ToR) switch the tier one. A number of racks together form a Performance Optimized Data center (POD) which are nowadays 20 or 40 ft. containers. The servers are usually connected via a 1 Gbps Ethernet cable to the ToR switch who are also connected with the same bandwidth to the second tier. The second tier is formed by layer-3 switches which on the one hand connect the racks within the containers and on the other hand interconnect the containers using currently 10 GE links. According to Kliazovich et al. [7], Equal Cost Multi-Path (ECMP) routing is used for load balancing. Typically, a two-tiered design can support between 5,000 to 8,000 hosts [8]. To reduce the number of links and thus the costs of the equipment for the two-tier architecture, the branches of the trees are usually oversubscribed by a factor of 1:2.5 to 1:8 [8].

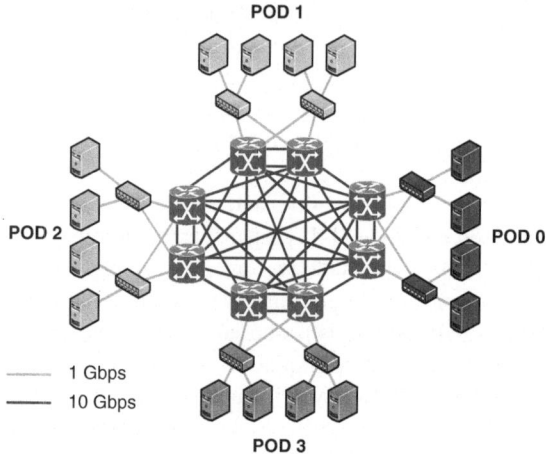

Fig. 1. Two-tier data center architecture

2.2   Three-Tier Architecture

The three-tier data center architecture is currently the most common architecture. It consists of three different layers, the access layer, the aggregation layer, and the core layer as shown in Figure 2. The aggregation layer facilitates the increase in the number of server nodes (more than 10,000 servers) while keeping inexpensive layer-2 switches in the access network for providing a loop-free topology. Similar to the two-tier architectures, the branches of the tree are oversubscribed and the highest levels of the tree can be oversubscribed by a factor of 1:80 to 1:240 [9]. The reason is that the three-tier architecture is often used for data processing such as the MapReduce algorithm. For this, the exchange of data is mostly kept within one rack and only one-tenth of the traffic is sent outside a rack. The three-tier architecture also normally uses ECMP for load balancing and as the maximum number of allowed ECMP paths is eight, a typical three-tier architecture consists of eight core switches. Figure 2 only shows two core switches. The current connection between the layers is similar to the two-tier architecture. However, it is intended to increase the link speed between the aggregation layer and the core layer to 40 GE or even 100 GE links [7].

2.3   DCell Architecture

The DCell data center architecture was developed to provide a scalable infrastructure and to be robust against server failures, link outages, or server-rack failures [10]. A DCell physical structure is a recursively defined architecture whose servers have to be equipped with multiple network ports. Each server is connected to other servers and to a mini switch, cf. Figure 3. In the example, $n = 4$ servers are connected to a switch, forming a level-0 DCell. According to Guo et al. [10], n should be chosen ≤ 8 to be able to use commodity 8-port switches with 1 Gbps or 10 Gbps per port. A level-1 DCell is constructed using $n + 1$ level-0 DCells, in our example 5 level-0 DCell form the level-1

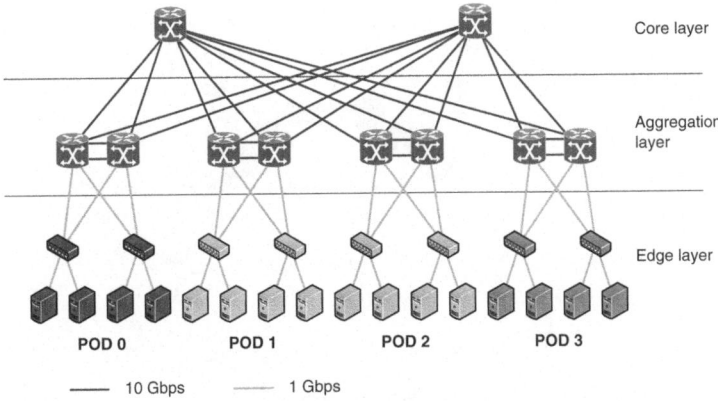

Fig. 2. Three-tier data center architecture

DCell. In order to connect the level-0 DCells, each DCell is connected to all other DCells with one link. A level-2 DCell and the level-k DCell are constructed the same way. Thus, the DCell architecture is a server-centric structure which uses commercial switches and the fewest number of switches of all presented data center architectures. However, the cabling complexity might prevent large deployments.

The goal of the DCell scheme is according to Guo et al. [10] to interconnect up to millions of servers. Thus, a global link-state routing scheme cannot be applied. Therefore, a new routing protocol is proposed, called DCell Fault-tolerant Routing (DFR) which is a decentralized touring solution. More information about the routing protocol can be found in Guo et al. [10].

2.4   BCube Architecture

BCube is similar to the DCell structure, just that the server-to-server connections are replaced by server-to-switch connections for faster processing [11]. Figure 4 shows a BCube$_k$ ($k = 1$) architecture with $n = 4$ servers per switch. From the figure we can see that the total number of servers is $N = n^{k+1}$ and each server has to be equipped with $k + 1$ ports. Each level has n^k switches and the total number of levels is $k + 1$. Similar to DCell and in contrast to the following fat-tree architecture, BCube is server-oriented and can use existing commercial Ethernet switches. To be able to fully utilize the multi-path structure of the BCube and to automatically load-balance the traffic, a BCube Source Routing (BSR) protocol is proposed by Guo et al. [11]. In the paper it is also shown that the BCube architecture is more robust against server and switch failures compared to the DCell architecture and the following fat-tree architecture. However, in contrast to the DCell architecture, the BCube architecture should mainly be used for server interconnection within a container. To create larger data center architectures with more than 2,500 server, another architecture is proposed which is called Modularized Data center Cube (MDCube) [3]. With MDCube, multiple BCubes are interconnected by using 10 Gbps interfaces of switches in BCube. The routing between the different containers is realized using single-path routing.

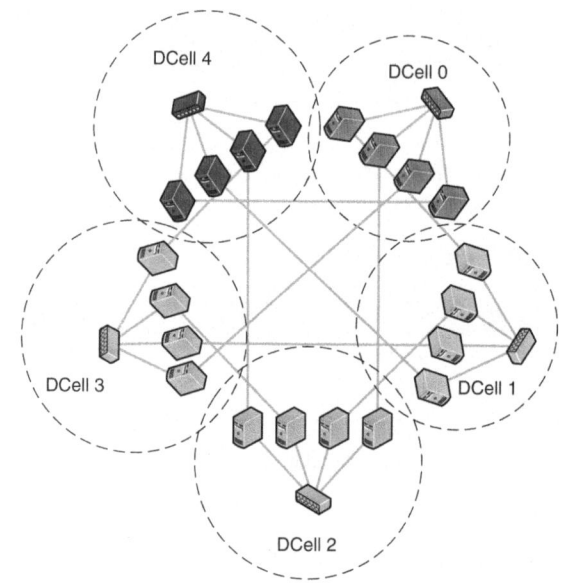

Fig. 3. DCell data center architecture

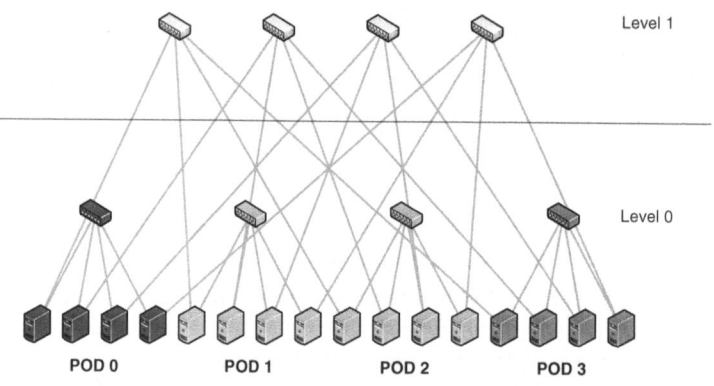

Fig. 4. BCube data center architecture

2.5 Fat-Tree Architecture

In contrast to the general three-tier topology and similar to the DCell and BCube architecture, a fat-tree topology uses commercial Ethernet switches [8, 12]. The fat-tree architecture was developed to reduce the oversubsciption ratio and to remove the single point of failures of the hierarchical architecture. As similar switches are used on all layers of the architecture, the costs for setting up a fat-tree data center can be kept low. The architecture is not achieving complete 1:1 oversubscription in reality, but offers rearrangeably non-blocking paths with full bandwidth. An example of a fat-tree data center architecture is shown in Figure 5. The figure shows a 4-ary fat-tree which is build up of $k = 4$ PODs, each containing two layers of $k/2$ switches.

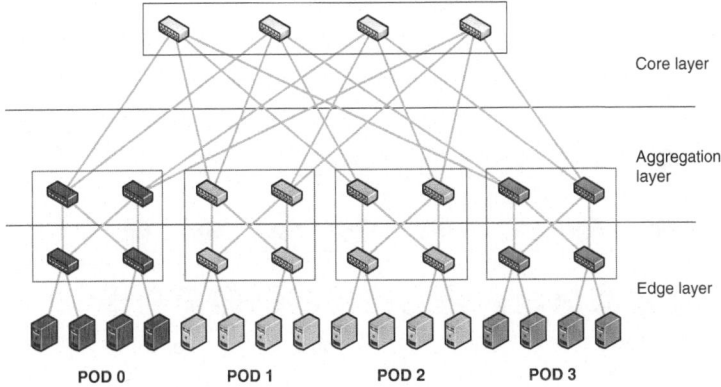

Fig. 5. Fat-tree data center architecture

The switches in the edge layer are connected to $k/2$ servers and the remaining ports of the edge switches are connected to the aggregation layer, cf. Figure 5. The core layer consists of $(k/2)^2$ k-port core switches where each of them is connected to each of the k PODs [8]. A fat-tree data center architecture built with k-port switches support $k^3/4$ servers. Thus, when using 48-port switches, up to 27,648 server can be supported. The example in Figure 5 shows that fat-tree is a switch-centric structure where the switches are concatenated. The VL2 architecture proposed by Greenberg et al. [9] is quite similar to fat-tree except that fewer cabling is needed. They claim that switch-to-switch links are faster than server-to-switch links and therefore use 1 Gbps links between server and switch and 10 Gbps links between the switches. By this, they reduce the number of cables required to implement the Clos. However, high-end intermediate switches are needed and thus, the trade-off made is the cost of those high-end switches.

2.6   Elastic-Tree Architecture

All the above mentioned mesh-like approaches help to be robust against failures by using more components and more paths which of course also increases the power consumption. However, although the number of traffic fluctuates during the day, the power consumption is fixed, see e.g. Google production data center [13]. Thus, Heller et al. [13] propose to reduce the power consumption by dynamically turning off switches and links that are not needed. The approach is called elastic-tree whose underlying topology is a fat-tree. Figure 6 shows an example of the elastic tree, where 7 switches are turned off compared to the normal fat-tree topology.

Using such energy-efficient data center architecture, it has to be ensured that the performance does not degrade, meaning that in case of high load, the switches should be able to start up almost immediately to enable multi-path transmissions. In addition, also in case of switch failure, the elastic-tree architecture has to immediately react to it. Taking these challenges into account, we will later see the effect on the overall power consumption.

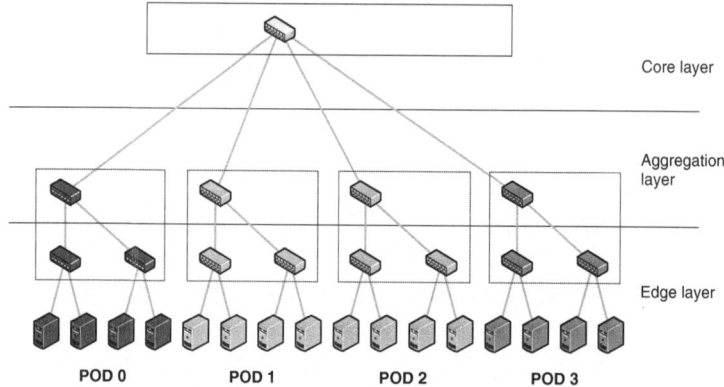

Fig. 6. Elastic-tree data center architecture

3   Evaluation Setup

To evaluate the power consumption of the six introduced data center architectures, we use the parameters shown in Table 1. The parameters were either measured ourselves, taken from published papers, or taken out of the handbook of the switches and routers. For the evaluation in the next section, we use these parameters and choose the required switch depending on the data center architecture as well as on the size of the data center. We scale the number of servers from one or a few hundred, depending on the architecture, to up to 70,000 servers. The evaluation of the power consumptions and the shares of the different parts responsible for the energy consumption is done using Matlab. In the next section, we show the results of our study.

Table 1. Parameters used for evaluation

	Consumption	**Reference**
server	145 Watt	HP ProLiant 2.13 GHz 10 GB RAM
cabling	0.4 Watt (1 Gbps) 6 Watt (10 Gbps)	[7]
linecard	5 Watt	[11]
COTS switch	145 Watt (48 port) 100 Watt (24 port) 13.4 Watt (16 port) 6 Watt (8 port)	NEC IP8800 NEC IP8800 D-Link DGS-1016D D-Link DGS-1008D
Core switch/ router	198 Watt (48 port) 3,500 Watt (128 port) 10,700 Watt (512 port)	HP A9508-V HP A12500

4   Performance Evaluation

Using the parameters described in the previous section, we first compare all data center power consumption values for a varying number of servers. The results are shown in Figure 7. The results show that the overall power consumption is quite similar, with only minor differences. The two-tier and three-tier architectures together with the BCube architecture have the lowest power consumption while the DCell architecture shows the worst performance. However, all architectures have a power consumption between 10 and 12 MWatt for 70,000 servers.

The similarity of the results rises to the suspicion that the servers are the main contributors of the overall power consumption. To underline this, we now take a look at the shares of the power consumers for the architectures. This is shown in Figure 8. The figure illustrates that more than 88 percent of the total power is consumed by the servers for all data center architectures.

The second largest consumer when using the DCell or the BCube architecture are the linecards. The reason for this huge amount of power consumption is that the servers are included in the switching process and that for each hierarchy level an additional linecard is needed within the server. For all other architectures, the switches are the second largest consumer. Surprising is that the two-tier and three-tier switches have a lower power consumption compared to the fat-tree switches although layer-3 core switches with a lot more power consumption are used. The reason is that the fat-tree architectures uses a lot more switches compared to the other two architectures to be resilient against network failures. Now that we know that the main power consumers are the servers, we can focus on the network equipment to see the differences of the architectures. Figure 9 shows these differences again for an increasing number of servers. For less than 18,000

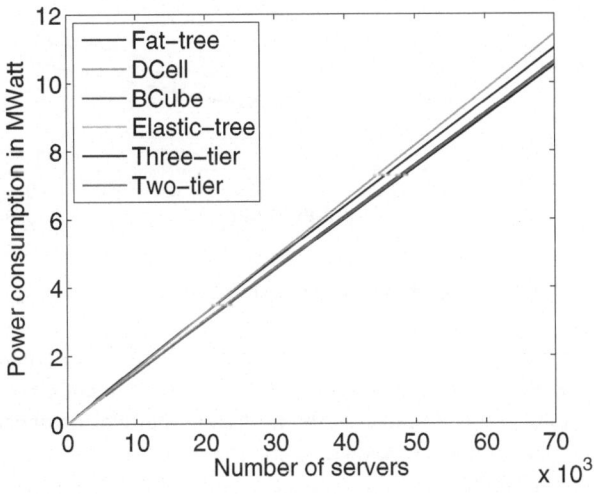

Fig. 7. Overall power consumption for different data center architectures

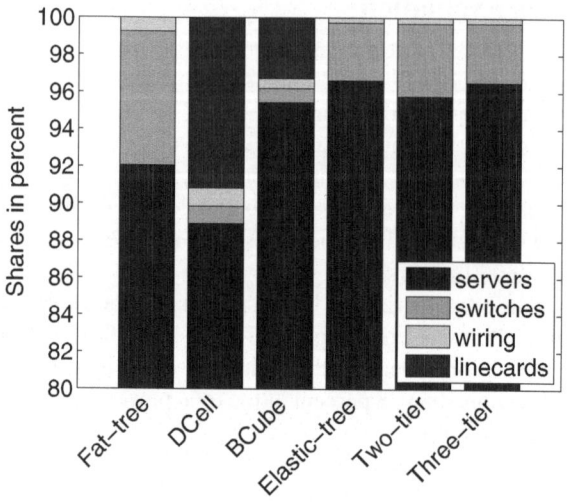

Fig. 8. Relative total power consumption

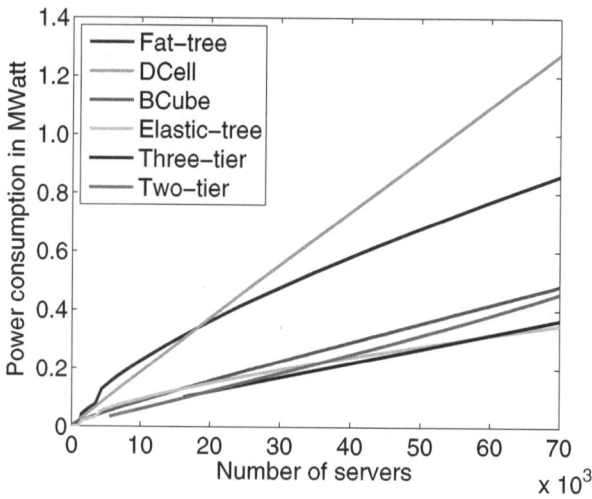

Fig. 9. Network power consumption

servers, the fat-tree architecture shows the worst performance but when increasing the number of servers, the power consumption of the DCell architecture overtakes the fat-tree power consumption. The reason is the increasing number of linecards within the servers. The best performance is shown by the three-tier and the elastic-tree architecture. Both power consumptions are less than one-third of the DCell power consumption for 70,000 servers. However, we have to keep in mind that the elastic-tree architecture uses COTS hardware while the three-tier architectures requires costly layer-3 switches.

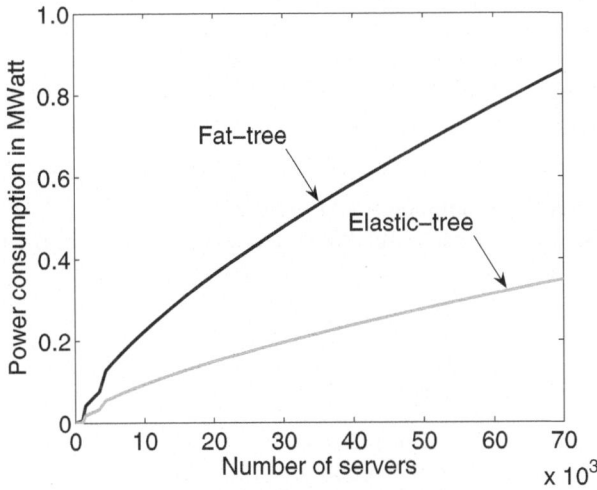

Fig. 10. Power savings of elastic-tree approach

Finally, we want to directly compare the fat-tree architecture with the elastic-tree approach as both use the same architecture, with the only difference that the elastic-tree approach switches off unused components to save energy. The direct comparison is shown in Figure 10.

It can easily be seen that the network equipment of the elastic-tree architecture consumes about half of the power compared to the fat-tree architecture. Thus, the potential for energy saving is tremendous just by turing off unused network equipment. However, in case of a network failure, the unused equipment has to be switched on as fast as possible to avoid data loss.

In addition to the network equipment, also the servers can be switched off when the load in the data center is low. In such a case, the jobs can be migrated to as few servers as possible, while the other are switched off. However, also here the startup time of the servers have to be taken into account and thus, there is always the trade-off between energy-efficiency and Quality of Service.

5   Conclusion

Although the servers in a data center consume most of the power, we showed in this paper that the power consumption of the network equipment should not be neglected. About 4% to 12% of the overall power consumption can be attributed to the networking hardware. Here, the three-tier architecture shows the best performance but uses the most costly hardware. However, the results in this paper illustrate that the total power consumption depends not only on the used data center architecture but also on the implemented energy saving mechanisms. For example, the fat-tree architecture - when used as proposed - consumes a lot of power due to the resilient paths to the servers.

When not used networking components are switched off, the power consumption can be reduced by about 60% as shown with the elastic-tree architecture.

In future work, we will implement the elastic-tree approach in real hardware and we want to consider also a possible server switch off. Therefore, we will have to consider the time needed for virtual machine migration as well as the time needed to switch a server on.

Acknowledgments. The authors would gratefully thank Michael Düser and Fritz-Joachim Westphal from Deutsche Telekom Laboratories for the fruitful discussions and support on this paper.

References

1. Facebook: Open Compute Project (2011),
   http://opencompute.org/
2. ONF: Open Networking Foundation (2011),
   http://www.opennetworkingfoundation.org/
3. Wu, H., Lu, G., Li, D., Guo, C., Zhang, Y.: MDCube: A high performance network structure for modular data center interconnection. In: Proceedings of the 5th International Conference on Emerging Networking Experiments and Technologies (CoNEXT), Rome, Italy, pp. 25–36 (2009)
4. Chen, K., Hu, C., Zhang, X., Zheng, K., Chen, Y., Vasilakos, A.V.: Survey on routing in data centers: Insights and future directions. IEEE Network 25(4), 6–10 (2011)
5. Poess, M., Nambiar, R.O.: Energy cost, the key challenge of today's data centers: A power conumption analysis of TPC-C results. VLDB Endowment 1(2), 1229–1240 (2008)
6. Gyarmati, L., Trinh, T.A.: How can architecture help to reduce energy consumption in data center networking. In: e-Energy 2010: Proceedings of the 1st International Conference on Energy-Efficient Computing and Networking, Passau, Germany, pp. 183–186 (2010)
7. Kliazovich, D., Bounvry, P., Audzevich, Y., Khan, S.U.: Greencloud: A packet-level simulator of energy-aware cloud computing data centers. In: IEEE Globecom, Miami, FL, USA (2010)
8. Al-Fares, M., Loukissas, A., Vahdat, A.: A scalable, commodity data center network architecture. In: SIGCOMM 2008: Proceedings of the ACM SIGCOMM 2008 Conference on Data Communication, Seattle, WA, USA, pp. 63–74 (2008)
9. Greenberg, A., Hamilton, J.R., Jain, N., Kandula, S., Kim, C., Lahiri, P., Maltz, D.A., Patel, P., Sengupta, S.: VL2: A scalable and flexible data center network. SIGCOMM Comput. Commun. Rev. 39(4), 51–62 (2009)
10. Guo, C., Wu, H., Tan, K., Shi, L., Zhang, Y., Lu, S.: DCell: A scalable and fault-tolerant network structure for data centers. SIGCOMM Comput. Commun. Rev. 38(4), 75–86 (2008)
11. Guo, C., Lu, G., Li, D., Wu, H., Zhang, X., Shi, Y., Tian, C., Zhang, Y., Lu, S.: BCube: a high performance, server-centric network architecture for modular data centers. In: SIGCOMM 2009: Proceedings of the ACM SIGCOMM 2009 Conference on Data Communication, Barcelona, Spain, pp. 63–74 (2009)
12. Mysore, R.N., Pamboris, A., Farrington, N., Huang, N., Miri, P., Radhakrishnan, S., Subramanya, V., Vahdat, A.: Portland: a scalable fault-tolerant layer 2 data center network fabric. SIGCOMM Comput. Commun. Rev. 39(4), 39–50 (2009)
13. Heller, B., Seetharaman, S., Mahadevan, P., Yiakoumis, Y., Sharma, P., Banerjee, S., McKeown, N.: Elastic tree: Saving energy in data center networks. In: 7th USENIX Symposium on Networked System Design and Implementation (NSDI), San Jose, CA, USA, pp. 249–264 (2010)

Gradient Optimisation
for Network Power Consumption

Erol Gelenbe and Christina Morfopoulou

Dept. of Electrical & Electronic Engineering
Imperial College, London SW7 2BT, UK
{e.gelenbe,c.morfopoulou}@imperial.ac.uk

Abstract. The purpose of this paper is to examine how a gradient-based algorithm that minimises a cost function that includes both quality of service (QoS) and power minimisation in wired networks can be used to improve energy savings with respect to shortest-path routing, as well as against a "smart" autonomic algorithm called EARP which uses adaptive reinforcement learning. Comparisons are conducted based on the same test-bed and identical network traffic. We assume that due to the need for network reliability and resilience we are not allowed to turn off routers and link drivers. We also assume that for QoS reasons (notably with regard to jitter and to avoid packet desequencing) we are not allowed to split traffic from the same flow into different paths. Under these assumptions and for the considered traffic, we observe that power consumed with the gradient-optimiser is a few percent to 10% smaller than that consumed using shortest-path routing or EARP. Since the magnitude of the savings is small, this suggests that further power savings may only be obtained if under-utilised equipment can be dynamically put to sleep or turned off.

1 Introduction

Much work has been devoted to power savings in wireless sensor networks where battery power can be crucial, including Topology Control [1,2,3] dynamically adjusting radio transmission power and hence range so as to preserve connectivity of each potential source-destination pair. In [4] it is indicated that the radio transceiver, which is the dominant energy consumer within a sensor, consumes almost the same amount of energy in transmit, receive and idle mode. In [5] energy efficient routing for ad hoc networks is presented using the Cognitive Packet Network (CPN) [6,7] with smart packets that improve QoS and energy savings. A recent comprehensive survey on energy efficiency in mobile networks can be found in [8]. Early work for wired networks [9] proposed traffic aggregation along a few routes, a modification of network topology by route adaptation and putting certain nodes and devices to sleep. A network-wide approach (coordinated sleeping) as well as a link layer approach (uncoordinated sleeping) were discussed and the possible effects on routing protocols were examined. In [10] an energy saving algorithm for Ethernet links using local data to make sleeping decisions was suggested, while powering components on/off in combination with

Joel J.P.C. Rodrigues et al.: (Eds.): GreeNets 2011, LNICST 51, pp. 125–134, 2012.
© Institute for Computer Sciences, Social Informatics and Telecommunications Engineering 2012

an offline multicommodity network-flow problem for traffic assignment was considered in [11]. An online technique was proposed [12] to spread load through multiple paths, based on a step-like model of power consumption as a function of the hardware's processing rate and the ability of nodes to automatically adjust their operating rate to their utilization. Rate-adaptation for individual links was examined in [13] based on the utilization and the link queuing delay, where traffic is sent out in bursts at the edge routers enabling other line cards to sleep between successive bursts. In [14] the authors select the active links and routers to minimize power consumption via simple heuristics that approximately solve a NP-hard problem. In [15] a case study based on specific backbone networks is discussed, and an estimate of the potential overall energy savings in the Internet is presented in [16]. In [17] the reduction of power consumption in wired networks in the presence of users' QoS constraints and experiments with dynamic traffic management in conjunction with the turning on/off of link drivers and/or routers is discussed, and using the Cognitive Packet Network (CPN) [18] routing protocol for energy awareness in conjunction to QoS is considered. Energy efficiency is examined for Cloud Computing in [19]. Power measurements of network components [11] indicate that the base system is the largest power consumer: it is best to minimize the number of chassis at a given point of presence (PoP) and maximize the number of cards per chassis. In [13] the impact of the hardware processing rate and traffic on power consumption yields measurements similar to those in Figure 1 from [20] for routers.

Here we apply a queuing theory based gradient method described in [21] for QoS and power minimisation in wired networks to improve upon (i) shortest-path routing and (ii) an experimental autonomic algorithm (EARP) [22] for QoS and power optimisation. The algorithm is limited to a single step of the gradient descent in order to provide fast computation, and the algorithm is initiated for (i) with the shortest path algorithm, and for (ii) with EARP. Comparisons are conducted using the same test-bed and the same network traffic. We assume that due to the need for network reliability and resilience we do not turn off routers and link drivers, and that for QoS reasons (notably with regard to jitter and packet desequencing) we do not split traffic from the same flow into different paths. Under these assumptions we observe that the energy consumed using the gradient-optimiser when it is started with known shortest-paths or with paths discovered by EARP is smaller by a few percent, and savings are greater when starting with paths provided by EARP which selects paths based on power optimisation. We note that even a few percent in power savings, scaled up to the power consumed in high-speed routers over long periods of time, can lead to significant economic and carbon dioxide (CO_2) savings in energy. The CO2 emissions obviously depend on whether the energy used is from nuclear, fossil, or renewable sources such as hydroelectric, wind or electrovoltaic. However energy savings will have at least the same proportion in CO2 savings, because lower energy use offers greater opportunity to run routers and link drivers using a combination of renewable and stored energy, and there is also less need for cooling using fans and air conditioning equipment.

2   Network Optimisation

Probability models have long been used to design and optimise computer systems and architectures [23]. Here we follow the same tradition by designing an algorithm that is based on a class of probability models for networks called G-networks [24,25] with multiple classes [26], that were initially inspired by biological neural systems [27]. This mathematical model includes the queues that form at routers and links due to the flow of payload packets, as well as the flow of control packets that are used to re-route traffic for power savings or QoS improvement. For lack of space, we only sketch the optimisation algorithm which can be found elsewhere [21]. The contribution of this paper is to evaluate how this algorithm will improve power consumption over well known shortest path routing and the smart adaptive algorithm EARP. The G-network model allows us to represent the fact that control packets used to optimise are also adding to congestion and power consumption, and the model includes their effect on performance, on the overhead that they induce, and on power consumption.

The network model that is used in the optimisation algorithm model has a set of N queues: router queues R or link queues L, $\mathbf{N} = \mathbf{R} \cup \mathbf{L}$. We use r and l to denote a router or link, $r \in \mathbf{R}$ and $l \in \mathbf{L}$. A traffic class k is a flow of packets between a source-destination pair (s, d), which travels on a path to destination. $\lambda(r, k)$ is the packet rate of user class k arriving from outside the network to router r and $\lambda(r, k) > 0$ only if r is the source node for class k. The probability that a packet of class k travels in one step from node i to node j is $P(i, k, j)$. We also have control traffic classes (r, k) acting on user class k at router r, and $\lambda^-(j, (r, k))$ is the rate at which such control packets may enter the network via router i. The probability that a control packet of class (r, k) travels from node i to j in one hop is $p((r, k), i, j)$. The control classes may also be virtual representations of rerouting decisions; in that case these "virtual packets" will not create traffic overhead but will generate computational overhead at the nodes where decisions are taken. The probability that a user of class k is directed from router r to neighbour j by a control packet is $Q(r, k, j)$. Links will have only one predecessor and successor, while routers may have one or more successors that are links. Note also that some models may abstract the existence of links, and just represent the manner in which routers are connected without detailing the links. The equations of the network model are:

$$\Lambda_R(r, k) = \lambda(r, k) + \sum_{l \in \mathbf{L}} q(l, k) P(l, k, r) \mu_l(l), \ r \in \mathbf{R}$$

$$\Lambda_L(l, k) = \sum_{r \in \mathbf{R}} [P(r, k, l) q(r, k) \mu_r(r, k)$$

$$+ \Lambda^-(r, (r, k)) q(r, k) Q(r, k, l)], \ l \in \mathbf{L}$$

$$q(r, k) = \frac{\Lambda_R(r, k)}{\mu_r(r, k) + \Lambda^-(r, (r, k))}, \ r \in \mathbf{R}$$

$$q(l, k) = \frac{\Lambda_L(l, k)}{\mu_l(l)}, \ l \in \mathbf{L},$$

where $\Lambda_R(r, k)$, $\Lambda_L(l, k)$, $q(r, k)$, $q(l, k)$ denote the total arrival rates to the routers and links, and the utilisation rate for the routers and links, for user traffic class k. The corresponding quantities for control traffic class (i, k) are given by:

$$\Lambda^-(j, (i, k)) = \lambda^-(j, (i, k)) + \sum_{l \in \mathbf{L}} p((i, k), l, j) c_L(l, (i, k)) \mu_l, \ i, j \in \mathbf{R}$$

$$= \sum_{r \in \mathbf{R}} p((i, k), r, j) c_R(r, (i, k)) \mu_r, \ i \in \mathbf{R}, j \in \mathbf{L}, \ i \neq r$$

$$c_L(l, (i, k)) = \frac{\sum_{r \in \mathbf{R}} p((i, k), r, l) c_R(r, (i, k)) \mu_r}{\mu_l}, l \in \mathbf{L}, i \in \mathbf{R}$$

$$c_R(r, (i, k)) = \frac{\lambda^-(r, (i, k)) + \sum_{l \in \mathbf{L}} p((i, k), l, r) c_L(l, (i, k)) \mu_l}{\mu_r}$$

The steady-state probability that router r is busy is

$$B_R(r) = \sum_{k \in \mathbf{U}} [q(r, k) + \sum_{i \in \mathbf{R}} c_R(r, (i, k))] \tag{1}$$

and the steady-state probability that link l is busy is

$$B_L(l) = \sum_{k \in \mathbf{U}} [q(l, k) + \sum_{i \in \mathbf{R}} c_L(l, (i, k))] \tag{2}$$

The average network delay for the user traffic is then:

$$T_N = \frac{1}{\Lambda_T^+} [\sum_{r \in R} \frac{B_R(r)}{1 - B_R(r)} + \sum_{l \in L} \frac{B_L(l)}{1 - B_L(l)}] \tag{3}$$

where $\Lambda_T^+ = \sum_{k \in \mathbf{U}} \sum_{r \in \mathbf{R}} \lambda(r, k)$ is the total user traffic entering the network. The cost f to be minimised via judicious routing will contain a function of the probabilities that nodes and links are busy, and the power consumption of the network:

$$P_N = \sum_{r \in R} P(r) + \sum_{l \in L} P(l) \tag{4}$$

where router power consumption is represented by

$$P(r) = \alpha_r + g_R(B_R(r)) + c_r \sum_{k \in \mathbf{U}} \Lambda_R^-(r, (r, k)), r \in \mathbf{R} \tag{5}$$

and α_r is the router's static power consumption, $c_r > 0$ is a constant, $g_R(.)$ is an increasing function of the packet processing rate, c_r is a proportionality constant related to router processing for re-routing control, and power consumption in a link is:

$$P(l) = \beta_l + g_L(B_L(l)), l \in \mathbf{L} \tag{6}$$

where β_r is the static power consumption and $g_L(B_L(l))$ is an increasing function. The routing optimisation algorithm [21] minimises f which in general includes both network power consumption and average user packet delay:

$$\frac{Use \ Q(i, k, j)}{To \ Minimise} \ f = P_N + c T_N \tag{7}$$

where $c \geq 0$ is a constant the establishes the relative importance of delay with respect to power.

2.1   Improving upon Shortest-Path Routing

All comparisons are carried out based on the 23 node test-bed shown in Figure 1 described in [22,20]. The measured power consumption of the fourteen nodes located on the circle are shown in the lower Figure 1. The service rates of the links are their 100 Mbps speeds. Also, virtual delays are added to service times so as to compensate for the short physical links that are used in the laboratory. The comparisons are carried out in the presence of flows travelling from source to destination with average traffic rates: Flow 1 (22,18) traffic rate 30kpps, Flow 2(23,19) traffic rate 10kpps, Flow 3 (21,17) traffic rate 20kpps. First, we apply the optimisation algorithm to the network started in a state where all flows follow the shortest path, and we focus on power ($c = 0$). We can select among seven alternative paths for each flow, and the optimisation yields a saving of 10 Watts, down from 1531 Watts, at the cost of an increase in average end-to-end delay of 3.3ms. Then we vary the input traffic of the 3 flows from 0.1 to 1.5 times their initial value, and the results in Figure 2 show a modest average power savings of 8.2 Watts while average packet delay increases. If we opt for both power and delay optimisation by adjusting c in (7) we can avoid the increase in average delay seen in Figure 3 and the average power savings is a modest but real 6.4

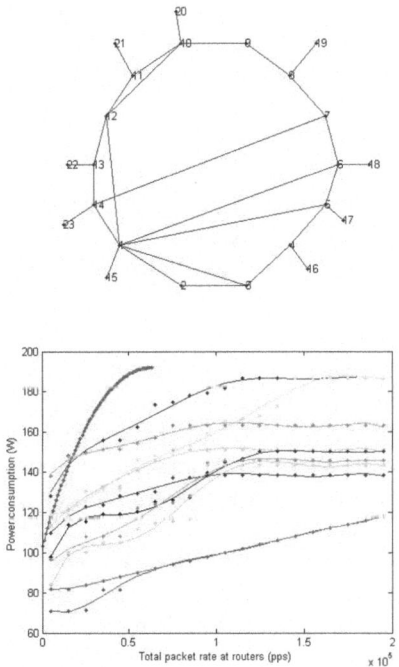

Fig. 1. Experimental network (top) and power profiles of 14 routers on the circle (bottom)

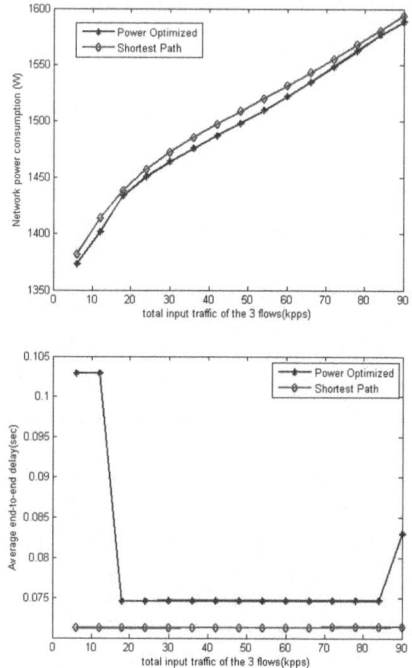

Fig. 2. Power consumption (top) and average end-to-end packet delay (bottom) against varying traffic load in Kpps (kilo-packets per second) for power-optimised versus shortest path routing

Watts. If we add another flow Flow 4 (20,16) at traffic rate 5kpps, and vary the traffic of the four flows from 0.1 to 1.5 times their nominal values, the average power savings increase to 13.1 Watts as shown in Figure 4.

2.2 Improving upon EARP

We now use the gradient based otimisation scheme to improve upon the on-line adaptive power and QoS protocol EARP [22]. EARP uses CPN to search for the paths that minimise a mixed QoS and power consumption criterion. The log files of observed paths for the three flows as they are generated by EARP are then used to initiate the optimisation algorithm. For every new path observed by EARP we run the gradient algorithm, and the outcome is shown in Figure 5 where we observe a significant power saving with respect to EARP (dashed lines represent the average values). In order to limit out of order packet arrival and jitter, we can change paths only when the path modification improves on the previous power consumption; we then have the greater power savings of Figure 6.

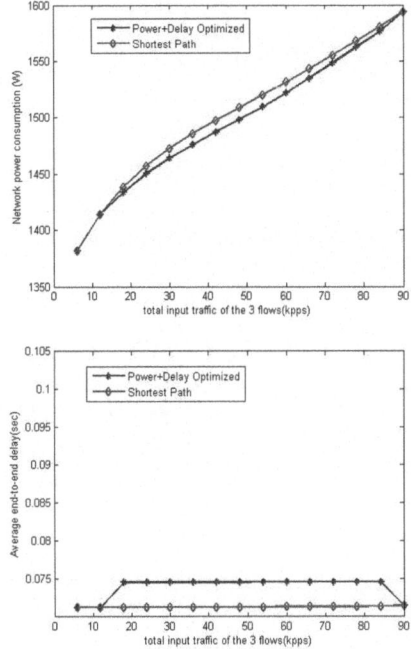

Fig. 3. Power consumption (above) and end-to-end delay with power and delay optimisation (below)

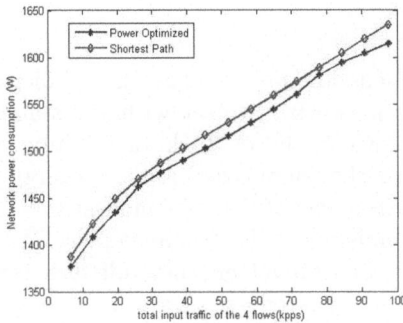

Fig. 4. Power consumption for proposed algorithm compared to shortest path with four flows and power optimisation

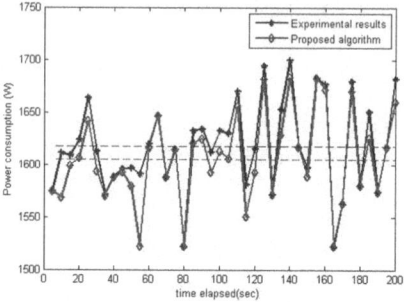

Fig. 5. Power consumption for the gradient algorithm (green) compared to power-based EARP [22] in blue

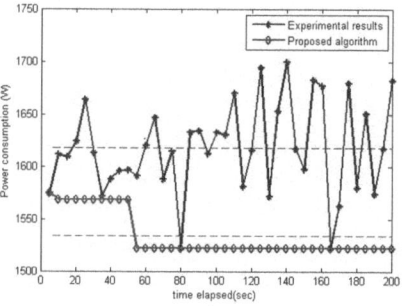

Fig. 6. Power consumption for proposed algorithm with memory

3   Conclusions

From the preceding discussion we see that the gradient based optimisation, started either with the network state set with the standard shortest-path algorithm for all of the flows in the network, or initially set using the self-ware EARP routing approach that adaptively reduces energy consumption, always brings some level of savings even if they are modest.

The impact on CO2 emissions will actually depend on the source of electrical energy, whether it is nuclear, or fossil, or renewable such as wind or electrovoltaic. However any energy savings will reflect at least at the same level in savings in the CO2 imprint of networks. We say "at least" because lower energy consumption would provide a better chance to run network nodes and link drivers for longer periods based on using renewable energy sources and reserve batteries when these sources are inactive. We note that if the experiments we have reported were scaled to a test-bed with high speed routers and large volumes of traffic, savings in energy costs and CO2 imprint can be substantial over long periods of time.

In previous work [21] we have shown that the algorithm we use in this paper is of time complexity $O(N^3)$, which would make it impractical for a large network. But in effect the algorithm can be simplified considerably because the gradient optimisation would in practice be only carried out for a limited number of nodes and a limited number of paths. For instance when we start with a shortest path, one will not look at all possible paths but rather work with other shortest paths or with paths which are at most one or two hops longer. Another simplification resides in the matrix inversion that leads to the $O(N^3)$ complexity. Rather than do a full matrix inversion, in most cases it may be sufficient to take the first two terms of the expansion of a matrix inversion $(I - W)^{-1} \approx (I + W + W^2)$ which can be faster than the matrix inversion. Also, any practical optimisation would be done in stages, working with successively smaller networks, or it may also be done hierarchically with a set of sub-network representations. We think that this part of the work can lead to many fruitful research options to study simplified algorithms and the impact that they will have on energy savings and QoS in the practical network.

Acknowledgment. The authors are grateful the research support of the Fit4Green EU FP7 Project co-funded under ICT Theme FP7-ICT-2009-4.

References

1. Rajaraman, R.: Topology control and routing in ad hoc networks: A survey. SIGACT News 33(2), 60–73 (2002)
2. Jia, X., Li, D., Du, D.: QoS topology control in ad hoc wireless networks. In: 23rd Annual Joint Conf. IEEE Computer and Comms. Societies, INFOCOM 2004, vol. 2, pp. 1264–1272 (March 2004)
3. Santi, P.: Topology control in wireless ad hoc and sensor networks. ACM Comput. Surv. 37(2), 164–194 (2005)
4. Boukerche, A., Cheng, X., Linus, J.: A performance evaluation of a novel energy-aware data-centric routing algorithm in wireless sensor networks. Wireless Networks 11(5), 619–635 (2005)
5. Gelenbe, E., Lent, R.: Power-aware ad hoc cognitive packet networks. Ad Hoc Networks 2(3), 205–216 (2004)
6. Gelenbe, E., Seref, E., Xu, Z.: Towards networks with intelligent packets. In: IEEE Conference on Tools for Artificial Intelligence, ICTAI 1999, IEEE Chicago, IL. IEEE (November 1999)
7. Gelenbe, E.: Steps towards self-aware networks. Communications of the ACM 52, 66–75 (2009)
8. Wang, X., Vasilakos, A.V., Chen, M., Liu, Y., Kwon, T.T.: A survey of green mobile networks: Opportunities and challenges. ACM/Springer Mobile Networks and Applications (2011), doi:10.1007/s11036-011-0316-4
9. Gupta, M., Singh, S.: Greening of the Internet. Computer Communication Review 33(4), 19–26 (2003)
10. Gupta, M., Singh, S.: Dynamic ethernet link shutdown for energy conservation on ethernet links. In: IEEE International Conference on Communications, ICC 2007, pp. 6156–6161 (June 2007)

11. Chabarek, J., Sommers, J., Barford, P., Estan, C., Tsiang, D., Wright, S.: Power awareness in network design and routing. In: IEEE INFOCOM 2008, pp. 457–465 (April 2008)
12. Vasic, N., Kostic, D.: Energy-aware traffic engineering. Technical report, EPFL (2008)
13. Nedevschi, S., Popa, L., Iannaccone, G., Ratnasamy, S., Wetherall, D.: Reducing network energy consumption via sleeping and rate-adaptation. In: NSDI 2008: Proc. 5th Symp. on Networked Systems Design and Implementation, pp. 323–336. USENIX Association, Berkeley (2008)
14. Chiaraviglio, L., Mellia, M., Neri, F.: Reducing power consumption in backbone networks. In: IEEE ICC 2009, pp. 1–6 (June 2009)
15. Chiaraviglio, L., Mellia, M., Neri, F.: Energy-aware backbone networks: A case study. In: IEEE International Conference on Communications Workshops, ICC Workshops 2009, pp. 1–5 (June 2009)
16. Chiaraviglio, L., Ciullo, D., Leonardi, E., Mellia, M.: How much can the Internet be greened? In: IEEE GLOBECOM Workshops, pp. 1–6 (December 2009)
17. Gelenbe, E., Silvestri, S.: Reducing power consumption in wired networks. In: 24th International Symposium on Computer and Information Sciences, ISCIS 2009, METU, September 14-16, pp. 292–297. IEEE Digital Library (2009)
18. Gelenbe, E.: Cognitive packet network (CPN). U.S. Patent 6,804,20 (October 11, 2004)
19. Berl, A., Gelenbe, E., Girolamo, M.D., Giuliani, G., Meer, H.D., Dang, M.Q., Pentikousis, K.: Energy-efficient cloud computing. The Computer Journal 53(7), 1045–1051 (2010)
20. Lent, R.: A sensor network to profile the electrical power consumption of computer networks. In: GLOBECOM 2010 Workshop on Green Computing and Communications. IEEE (December 2010)
21. Gelenbe, E., Morfopoulou, C.: A framework for energy-aware routing in packet networks. The Computer Journal 54(6), 850–859 (2011)
22. Gelenbe, E., Mahmoodi, T.: Energy-aware routing in the cognitive packet network. In: International Conf. on Smart Grids, Green Communications and IT Energy-Aware Technologies - Energy 2011 (May 2011)
23. Gelenbe, E.: A unified approach to the evaluation of a class of replacement algorithms. IEEE Transactions on Computers 22(6), 611–618 (1973)
24. Gelenbe, E.: G-networks with signals and batch removal. Probability in the Engineering and Informational Sciences 7, 335–342 (1993)
25. Gelenbe, E.: G-networks with instantaneous customer movement. Journal of Applied Probability 30(3), 742–748 (1993)
26. Gelenbe, E., Labed, A.: G-networks with multiple classes of signals and positive customers. European Journal of Operations Research 108(2), 293–305 (1998)
27. Gelenbe, E., Stafylopatis, A.: Global behaviour of homogenous random systems. Applied Mathmetical Modelling 15(10), 534–541 (1991)

On Multipath Transmission Scheduling in Cognitive Radio Mesh Networks

Brendan Mumey[1], Xia Zhao[2], Jian Tang[3], and Richard Wolff[2]

[1] Department of Computer Science
Montana State University, Bozeman, MT 59717
mumey@cs.montana.edu
[2] Department of Electrical and Computer Engineering
Montana State University, Bozeman, MT 59717
[3] Department of Electrical Engineering and Computer Science
Syracuse University, Syracuse, NY 13244

Abstract. Nodes in a cognitive radio mesh network comprised of secondary users may select from a set of available channels provided they do not interfere with primary users. This ability can improve overall network performance but introduces the question of how best to use these channels. Given a routing multipath M, we would like to choose which channels each link in M should use and a corresponding transmission schedule so as to maximize the end-to-end data flow rate (throughput) supported by the entire multipath. This problem is relevant to applications such as streaming video or data where a connection may be long lasting and require a high constant throughput as well as providing robust, high-speed communications in wireless mesh networks deployed in rural environments, where there are significant amounts of spectrum available for secondary use. Better transmission scheduling can lead to improved network efficiency and less network resource consumption, e.g. energy-use. The problem is hard to due the presence of both intra-flow and inter-flow interference. In this paper, we develop a new polynomial time constant-factor approximation algorithm for this problem. We also present an effective heuristic method for finding effective multipath routes. It has been shown by simulation results that the end-to-end throughput given by the proposed algorithms provide nearly twice the throughput of single path routes and that the schedules generated are close to optimal.

Keywords: Wireless mesh networks, cognitive radios, multipath scheduling, channel assignment, interference

1 Introduction

Wireless Mesh Networks (WMNs) are considered an economical method of providing robust, high-speed backbone infrastructure and broadband Internet access in large rural areas [1]. The mesh topology offers the advantages of alternative route selection to assure throughput and quality of service (QoS) requirements under dynamic load conditions. As aggregate traffic volume can be substantial on backbone links converging on gateways and servers, considerations of

Joel J.P.C. Rodrigues et al.: (Eds.): GreeNets 2011, LNICST 51, pp. 135–147, 2012.
© Institute for Computer Sciences, Social Informatics and Telecommunications Engineering 2012

transmission scheduling, path selection and topology control are essential to assure that a WMN can meet the QoS and throughput requirements of end-users' real-time multimedia applications. Furthermore, range considerations and propagation characteristics demand careful attention to interference.

Spectrum is the most precious resource for wireless networks. Licensed spectrum is currently managed on a static and non-preemptive basis, where the license holder has exclusive use of the designated frequencies in a geographic area. Spectrum that is licensed but unused is not available to others on a demand basis. Unlicensed spectrum is used on a non-exclusive basis, meaning that is available to all users on an equal basis, and rules of etiquette are required to assure that users can coexist on selected frequencies in a particular area. Over the past few years, the world has experienced a very rapid proliferation of wireless devices operating in both licensed and unlicensed spectrum. Certain unlicensed parts of the spectrum, such as the 2.4GHz band and the 5GHz band, are heavily used by various wireless devices, resulting in serious interference and poor network performance. There is still a significant amount of spectrum that remains under-utilized or even not utilized at all in the licensed spectrum bands, which has been shown by recent studies and experiments [2]. Ideally, these fallow portions of spectrum could be used on a secondary, or pre-emptive basis to alleviate the congestion and meet the growing demands of wireless applications. Such blocks of spectrum, sometimes deemed as *white spaces*, often appear in the broadcast television bands, where the licensees are migrating their services to cable and satellite distribution, and in rural areas, where broadcast television is in very limited use. Therefore, the traditional static licensed spectrum allocation approach does not efficiently manage the spectrum access any longer. Emerging cognitive radios enable dynamic spectrum access. With cognitive radios, unlicensed wireless users (a.k.a secondary users) can sense and access the under-utilized licensed or unlicensed spectrum bands opportunistically as long as the licensed wireless users (a.k.a primary users) in these spectrum bands are not disrupted [2]. In this way, interference can be avoided and network capacity, QoS and robustness can be significantly improved.

Cognitive radios are desirable for a WMN in which a large volume of traffic is expected to be delivered since they are able to utilize available spectrum more efficiently than conventional, static channel assignment methods and therefore improve network capacity significantly [2]. However, they introduce additional complexities to bandwidth allocation. With cognitive radios, each node can access a set of available spectrum bands which may span a wide range of frequencies. Each spectrum band may be divided into channels, and the channel widths may vary from band to band. Different spectrum bands can support quite different transmission ranges and data rates, both of which have a significant impact on resource allocation and interference effects.

In this paper, we consider the problem of multipath scheduling: we are given a multipath M from a source to a destination and similarly must create a schedule for each link that maximizes the end-to-end throughput. This problem is relevant to applications such as real-time streaming video or data where a connection may be long lasting and require a high constant throughput. Better transmission scheduling can lead to improved network efficiency and less network resource consumption, e.g. energy-use. This work is different from most previous works on transmission scheduling which usually deal with the problem of scheduling a set of links for link-layer throughput maximization. Here, we focus on end-to-end performance, and consider the problem of allocating resources (timeslots, channels) along a multi-hop routing path or multipath, which is a very hard problem due to the constraints related to intra-flow interference (interference among links belonging to a common flow) and inter-flow interference (interference among links belonging to different flows) [21]. Previously, we presented a polynomial time constant-factor approximation algorithm for the path scheduling problem [12]. In this work, we generalize the approach to provide a constant factor approximation algorithm to create transmission schedules for multipaths that can further improve the achievable end-to-end transmission rate over single path routes. The approximation ratio is $\frac{1}{(\Delta_1+1)(\Delta_2+1)}$ where Δ_1 is the maximum degree of a vertex the multipath to be scheduled and Δ_2 is the maximum degree of vertex found in certain subgraphs of a flow conflict-graph. In addition, we also present an effective heuristic routing algorithm to find multipaths that can lead to high end-to-end throughput.

The rest of the paper is organized as follows. We discuss related work in Section 2. We formally define the system model in Section 3. In Section 4, we describe the formal problem and present our proposed multipath scheduling algorithm. In Section 5, we describe a routing heuristic for finding multipaths and then present numerical results in Section 6. Finally, the paper is concluded in Section 7.

2   Related Work

Cognitive radio wireless networks have recently received extensive attention. In [23], the authors derived optimal and suboptimal distributed strategies for the secondary users to decide which channels to sense and access with the objective of throughput maximization under a Partially Observable Markov Decision Process (POMDP) framework. In [24], Zheng et al. developed a graph-theoretic model to characterize the spectrum access problem and devised multiple heuristic algorithms to find high throughput and fair solutions. In [22], the concept of a time-spectrum block was introduced to model spectrum reservation, and a centralized and a distributed protocol were presented to allocate such blocks for cognitive radio users. Tang et al. introduced a graph model to characterize the impact of interference and proposed joint scheduling and spectrum allocation algorithms for fair spectrum sharing based on it in [16]. In [5], a distributed spectrum allocation scheme based on local bargaining was proposed for wireless ad-hoc networks with cognitive radios.

Cross-layer schemes have also been proposed for cognitive radio wireless networks. In [7], the authors proposed the Asynchronous Distributed Pricing (ADP) scheme to solve a joint spectrum allocation and power assignment problem.

In [18], Wang *et al.* presented a joint power and channel allocation scheme that uses a distributed pricing strategy to improve the network performance. In [20], a novel layered graph was proposed to model spectrum access opportunities, and was used to develop joint spectrum allocation and routing algorithms. In [19], the authors presented distributed algorithms for joint spectrum allocation, power control, routing and congestion control. A mixed integer non-linear programming based algorithm was presented to solve a joint spectrum allocation, scheduling and routing problem in [6]. A distributed algorithm was presented in [15] to solve a joint power control, scheduling and routing problem with the objective of maximizing data rates for a set of user communication sessions. In [17], a PTAS is presented for a more general maximum multiflow scheduling problem (maximize the total flow of a set of commodities with no specific routing path) and several constant-factor approximations are given for special cases. This paper also points out some errors in previous work on that problem. In [9], Karnik *et al.* proposed an optimal flow scheduling for multihop networks in the more general SINR model for interference. Their approach was based on solving a linear program (of potentially exponential size). More recently, in [10,11], the authors explore the use of column generation methods for improving the efficiency of finding optimal multiflow schedules in the SINR model.

The differences between this work and previous works are summarized as follows: 1) We consider a channel assignment and scheduling problem for a given routing multipath in cognitive radio networks with heterogeneous channels with the objective of maximizing end-to-end throughput, which is different from those works addressing link layer (single-hop) throughput such as [5,7,16,18,22,23,24]. 2) We propose a provably good algorithm to solve the formulated problem. However, many related works (such as [5,7,18,20]) only presented heuristic algorithms which cannot provide any performance guarantees. 3) While our link interference model is more idealized than SINR-based models, our multipath scheduling algorithm runs in low-degree polynomial time in contrast to the potentially exponential time methods proposed for SINR-based scheduling [9,10,11].

3   System Model

We consider a wireless mesh backbone network with static mesh routers and study the problem of scheduling transmissions along a path from a source node to a destination node so as to maximize the end-to-end throughput. We focus on a dynamic setting where source-destination connection requests arrive intermittently and once a routing path is established for a request, a schedule must be quickly constructed. Each establish source-destination flow exists for some time and reduces the availability of network resources for subsequent routing requests. In addition, similar to [4,16], a spectrum server is assumed to manage the spectrum allocation and scheduling in the network. It can collect channel availability information from the FCC's database and computes a spectrum allocation and scheduling solution using the proposed algorithm and broadcasts it to all the users at the beginning of each scheduling period. All the users can then access the spectrum according to the received solution.

We define our assumptions about the parameters of the cognitive radio network: Let m be the number of channels available in the network. In general, each link e_i will have only a subset of these channels available at any given time. This can be due to interference, the link distance being greater than the transmission range, or that channel being already in use on that link. We will also assume that each available channel j on link e_i has an associated bit rate $b_{e,j} \geq 0$. This bit rate can depend on the link distance and other factors.

We assume that communication in the network is done using synchronized transmission frames of a fixed length L. For simplicity we assume a slightly idealized case where the transmission frame is infinitely divisible, although a simple rounding scheme can be employed to produce an integer time slot schedule [12]. Let C_e be the set of channels available to link e during the current frame. We define a variable $f_{e,j} \geq 0$ to indicate the flow amount allocated on the link e on channel j, where $j \in C_e$. A link flow $f_{e,j}$ is active if it is positive. An active link flow f must be scheduled at some point during the frame. We assume that a scheduled link flow $f_{e,j}$ occupies a single continuous interval $[s_{e,j}, s_{e,j} + f_{e,j}) \subset [0, L)$, where $s_{e,j}$ indicates the starting time for the link flow.

We adopt the following simple interference model. We assume that there is an interference distance R_j for each channel j such that a link $e = (u, v)$ *interferes* with another link $e' = (u', v')$ on channel j if and only if $|u - v'| \leq R_j$ or $|u' - v| \leq R_j$. We will also consider that the nodes in question are half-duplex. This means that nodes cannot simultaneously transmit and receive. The duplexing and interference constraints impose conditions on which link flows can be active at the same time. We summarize these conditions in a well-known *conflict graph*, $G_c = (V_c, E_c)$, where the vertices V_c are the link flow variables $f_{e,j}$ and the edges (undirected) indicate which pairs of link flows that cannot be scheduled simultaneously due to interference or duplexing constraints. For a transmission schedule to be *valid*, it must not contain any pair of conflicting scheduled link flows at any time; i.e. for any two active link flows $f_{e,j}$ and $f_{e',k}$, $(f_{e,j}, f_{e',k}) \in E_c \Rightarrow [s_{e,j}, s_{e,j} + f_{e,j}) \cap [s_{e',k}, s_{e',k} + f_{e',k}) = \emptyset$.

4   Multipath Scheduling

In this section, we first formalize the problem considered and then present an algorithm to solve the problem.

A multipath $M = (V_M, E_M)$ from s to t is a subgraph of G such that $s, t \in V_M$ and for any $v \in V_M$, there exists a simple path in M from s to t that goes through v as an intermediate node.

Let $v_{in} \subset E_M$ be the set of incoming edges to $v \in V_M$ and let v_{out} be the set of outgoing edges from v. The scheduled bit flow entering any interior vertex on the path is equal to the bit flow leaving that vertex. This leads to the following constraint:

$$\sum_{e \in v_{in}} \sum_j b_{e,j} f_{e,j} = \sum_{e \in v_{out}} \sum_j b_{e,j} f_{e,j}; \ \forall v \in V_M \setminus \{s, t\} \quad (1)$$

We are interested in the following optimization problem:

Definition 1. MaxFlow-Multipath: *Find a valid transmission schedule for the links in E_M that maximizes the total bit flow $F = \sum_{e \in t_{in}} \sum_j b_{e,j} f_{e,j}$ from s to t using links belonging to M, subject to (1).*

We note that this formulation is very similar to other multicommodity network flow formulations [9,17] but we view the problem in a restricted sense where M is a relatively small subset of all possible links in the network and each link in M belongs to a path from s to t.

The general algorithm approach will be to use graph coloring to identify individual link flows that can be scheduled simultaneously and then use linear programming to determine optimal link flow values and build a transmission schedule for the active link flows that maximizes the end-to-end throughput. The pseudocode is given in Algorithm 1.

Algorithm 1. Multipath-Schedule

Step 1 For the input multipath $M = (V_M, E_M)$, first color the links E_M in duplex-only conflict graph (two links conflict if they share an endpoint). Suppose D *duplex colors* are used.

Step 2 For each channel j and duplex color d, color G_c^{dj} using a simple greedy algorithm. Suppose g^{dj} colors are used.

Step 3 Associate *color variables* $o_1^{dj}, \ldots o_{g^{dj}}^{dj}$ to the colors used in coloring G_c^{dj}.

Step 4 Solve the following linear program (LP):
1. $x_d \geq 0, o_k^{dj} \geq 0;\ \forall d, j, k$
2. $\sum_{k=1}^{g^{dj}} o_k^{dj} \leq x_d;\ \forall j, d$
3. $\sum_d x_d = L$
4. Each link flow $f_{e,j}$ was given a color with associated variable o_k^{dj} in Step 1. Add the constraint $0 \leq f_{e,j} \leq o_k^{dj}$ to the LP.
5. Include the conservation-of-flow constraint given by (1).
6. Maximize $F = \sum_{e \in t_{in}} \sum_j b_{e,j} f_{e,j}$

Step 5 For each channel j and duplex color d, we define the starting times as follows: Let $s_1^{dj} = 0$ and $s_k^{dj} = s_{k-1}^{dj} + o_{k-1}^{dj}$ for $1 < k \leq g_{dj}$.

Step 6 Create a schedule S for the time frame with the following rule: A link flow $f_{e,j}$ associated with color variable o_k^{dj} will be active in the interval $[\sum_{i=0}^{d-1} x_i + s_k^{dj}, \sum_{i=0}^{d-1} x_i + s_k^{dj} + f_{e,j})$.

The main idea of the algorithm is to first use a graph coloring approach to on the conflict graph restricted to duplexing constraints only. We note that for a simple transmission path P, this requires two colors and decomposes that graph into odd and even links along the path. For a more general multipath M, we first color the vertices of M so that the endpoints of each link are given different colors. We refer to these colors as *duplex colors*. If $e = (u, v)$ is a link in M, then we consider e to have the same duplex color as u. In this way, the links (and their associate flows) also receive a duplex color. Suppose that D duplex colors are used; this paritions the links of M into D groups. In order to prevent duplexing conflicts, we will subdivide the frame into D separate intervals and only schedule links in the interval corresponding to their duplex color.

To ensure that for each channel, all conflicting link flows on that channel are scheduled at different times, will use some additional graph coloring. In particular, for each duplex color d and channel j, we consider the subgraph G_c^{dj} of the conflict graph G_c, consisted of only those flows $f_{e,j}$ on channel j for which e has duplex color d. We color each of the subgraphs G_c^{dj} separately and further divide the portion of the frame devoted to duplex color d into non-overlapping intervals for scheduling the link flows of each color (the intervals for different colors can overlap). In Step 3, the algorithm solves a linear program (LP) to find the optimal link flows subject to the color interval conditions. Steps 4 and 5 create the frame schedule from the LP solution.

While the algorithm as presented is centralized, in principle it is possible to create a distributed implementation for it. The algorithms requires performing distributed graph coloring of interference graphs and also a distributed approach to linear programming. There exist distributed algorithms to both of these problems that require only local sharing of information [3,14].

4.1   Analysis

It is clear by construction that this transmission schedule is valid, since no conflicting links are scheduled at the same time. Let F_S be the path bit flow obtained by the schedule created by Algorithm 1 and let F_{S^*} be the path bit flow obtained by an optimal schedule S^*.

Definition 2. *We say a transmission schedule is* duplex equal *if and only if link flows with duplex color $d \in \{1, \ldots, D\}$ are scheduled in the frame interval $[(d-1)\frac{L}{D}, d\frac{L}{D})$.*

Lemma 1. *Let S^{*de} be an optimal duplex-equal schedule for the path P with associated bit flow $F_{S^{*de}}$. Then $F_{S^{*de}} \geq \frac{1}{D}F_{S^*}$.*

Proof. A simple way to see this is to take the schedule S^* and scale it by $1/D$ to create a schedule for the half-frame $[0, L/D)$. Place a copy of the scaled S^* in each interval $[(d-1)\frac{L}{D}, d\frac{L}{D})$ and delete any link flows in this interval that do not have duplex color d. The resulting schedule is now duplex-equal with bit flow value $\frac{1}{D}F_{S^*}$, so the total flow for optimal duplex-equal schedule will be at least this value.

It is well-known that the greedy coloring algorithm used in Steps 1 and 2 provides a coloring that uses $\Delta(G) + 1$ colors, where $\Delta(G)$ is the maximum degree of a vertex in the input graph G. Let $\Delta_1 = \Delta(M)$, the maximum degree of a vertex in the multipath M and let $\Delta_2 = \max_{d,j} \Delta(G_c^{dj})$. Then for all d, j, $g^{dj}, h_j \leq \Delta_2 + 1$.

Lemma 2. $F_S \geq \frac{1}{\Delta_2 + 1}F_{S^{*de}}$

Proof. Since S^{*de} is a duplex-equal schedule, each link flow $f_{e,j}$ will satisfy $f_{e,j} \leq L/D$. Scale each link flow in S^{*de} by $\frac{1}{\Delta_2 + 1}$. The resulting link flows now satisfy

$0 \le f_{e,j} \le \frac{L}{D(\Delta_2+1)}$. Letting $x_d = L/D$ for each d and $o_k^{dj} = \frac{L}{D(\Delta_2+1)}$ for all d, j yields a feasible solution to the LP in Algorithm 1 with total bit flow $\frac{1}{\Delta_2+1}F_{S^{*de}}$. Since F_S is an optimal solution for the same LP, it follows that F_S will be at least $\frac{1}{\Delta_2+1}F_{S^{*de}}$.

We next observe the running time of Algorithm 1 is polynomial: Step 1 and 2 invokes a standard greedy coloring algorithm that runs in linear time in the size of the graphs colored. Steps 3 and 4 solve a linear program with $O(|V_c|)$ variables and constraints (note: the number of color variables is bounded by $|V_c|$ since giving each link flow its own unique color is a trivially valid coloring). This can be solved in $O(|V_c|^{3.5}L)$ using Karmarker's algorithm (L is the number of bits used to represent the input). Steps 5 and 6 require $O(|V_c|)$ time to create a valid link flow schedule. This leads to the following theorem.

Theorem 1. *Algorithm 1 is a $\frac{1}{(\Delta_1+1)(\Delta_2+1)}$-approximation algorithm running in polynomial time.*

Proof. From Lemmas 1 and 2 and the fact that $D \le \Delta_1 + 1$, Algorithm 1 produces a schedule S that satisfies

$$\begin{aligned}
F_S &\ge \frac{1}{\Delta_2+1}F_{S^{*de}} \\
&\ge \frac{1}{D(\Delta_2+1)}F_{S^*} \\
&\ge \frac{1}{(\Delta_1+1)(\Delta_2+1)}F_{S^*}.
\end{aligned}$$

5 Finding Multipath Routes

We present a simple multipath construction heuristic based on performing a depth first search (DFS) in G starting from the source node s. As subpaths to the destination node t are discovered, they are added to the multigraph M. A heuristic that we employ is to find a multipath that has chromatic index of 2; meaning that the nodes v in M can be assigned a parity (odd or even) and that all links in M have endpoints with opposite parities. This ensures that the number of duplex colors needed in Step 1 of Algorithm 1 is 2. The source vertex s is assigned odd parity. During DFS from a vertex u, if edge (u, v) is followed and v is already part of M, a check is made to see that u and v have opposite parities; if not that branch of the search fails and a different outgoing edge from u must be tried (if one exists). If u and v have opposite parities then the new subpath reaching v is added to M. This ensures the constructed multipath M has chromatic index 2. If the search path fails to reach t or a compatible existing path to t, the algorithm backtracks (clearing parity assignments as it goes back). A key decision to make while performing this search is the order in which outgoing edges are explored from any intermediate vertex u. This effects which subpaths get added to M. We use a simple heuristic rule to rank outgoing edges $e = (u, v)$ based on their capacity as well as their direction relative to the destination vertex v. We define the link capacity $c(e)$ as follows:

$$c(e) = \sum_{j \in C_e} b_{e,j}. \tag{2}$$

The link capacity provides an upper bound on the bit flow rate achievable by link e ignoring intra-path interference. Let θ_e be the angle between (u, v) (considering the link as vector) and the vector (u, t). Then the rank of an edge is defined by

$$r(e) = c(e) \cdot \cos(\theta_e). \tag{3}$$

We presort the adjacency lists of outgoing edges for each vertex u into decreasing order of $r()$-value and then conduct the depth-first search described above from the source vertex s. Once a DFS branch reaches t, that subpath is added to M. This continues until no new subpaths are discovered and DFS terminates. At this point, the multipath M is fully constructed.

6   Numerical Results

In the simulation, we used the DFS-based algorithm from Section 5 to compute multipath routes. As a benchmark, we also considered shortest path routes and another path routing approach that attempts to find a routing path whose links all have high estimated capacity as defined by (2). We define the *bottleneck capacity* of a path P to be $c(P) = \min_{e \in P} c(e)$. Our goal is to find a path P that maximizes $c(P)$. This is a well-known problem that can be efficiently solved by computing a minimum spanning tree T on the network graph using an edge weight function $w(e) = -c(e)$. The unique path in T from s to t will have maximum bottleneck capacity. For the path routes, we used our existing path routing solution proposed in [12]. In order to estimate how close to optimal the schedules are in practice, we also computed an upper bound on the optimal transmission schedule following the approach described in [12].

All of our numerical results were gathered using a random network created by placing 50 stationary nodes at random locations in a 50 km × 50 km grid. For our experiments, we assumed there was a maximum of 13 channels available per frequency band and that the link throughput for each channel was the maximum available given the link distance and frequency used. Three widely spaced frequency bands were chosen, typical of bands available for licensed and unlicensed operation in different regions of the world. The bands exhibit widely ranging propagation, transmission range and usage characteristics, highlighting the potential value of cognition in transmission scheduling. Tables 1 and 2 summarize our assumptions about the transmission rates and interference ranges of

Table 1. Maximum transmission distances by frequency and data rate

Transmission rate	700 Mhz	2400 Mhz	5800 Mhz
45 Mbps	15.4 km	4.5 km	1.8 km
40 Mbps	18.4 km	5.3 km	2.2 km
30 Mbps	30 km	8.6 km	3.6 km
20 Mbps	41 km	11.8 km	4.9 km
10 Mbps	68 km	20 km	8.2 km

Table 2. Interference ranges by frequency

Frequency	Interference range
700 Mhz	30.8 km
2400 Mhz	9 km
5800 Mhz	3.6 km

each frequency. These values are based on a scenario where each node transmits at 1W with a 2dBi antenna and the receiving mode antenna has a gain of 2dBi. The channel bandwidth is 10 MHz and the receiver noise figure is 5dB, and implementation losses of 3dB are assumed for each link. Path loss is calculated using line of sight and free space characteristics. Typical 802.16 adaptive modulation and coding parameters performance parameters are used to estimate the throughput achievable as a function of CNR (carrier to noise ratio), and are then translated into the allowable path loss threshold. The maximum channel transmission rate is a function of distance and frequency (at lower frequency, the maximum distance for a given transmission rate will be greater).

We also varied the number of primary users in the network (not shown). The number of primary users was set to be one third of the number of available channels used in the each particular scenario. Primary users were placed at random locations and assigned a random channel. This channel was then made unavailable to any link within the interference range of the primary user. The frame length L was set to 1 second and each frame was divided into 100 time slots.

6.1   Scenario 1: Performance on Random Connection Requests

In this scenario, 10 connection source-destination pairs were randomly created and routing paths and multipaths were found. The number of channels available in the network was set to 15 (5 from each frequency band). Each connection lasts 200 seconds. The transmission starting times for each path are staggered by 100 seconds, e.g. Connection 1 is active in the time interval $[0, 200]$, Connection 2 is active in the time interval $[100, 300]$, etc. The results are shown in Figure 1. We compared the performance of the schedules obtained to upper bounds on the optimal schedule performance. We found that, on average, the schedules obtained for shortest path routing, bottleneck path routing and multipath routing were, respectively, within 97%, 94%, and 86% of optimal.

6.2   Scenario 2: Varying the Number of Channels Available

In this scenario, the number of channels available to secondary users was varied from 9 to 39 in increments of 6 (chosen equally from each frequency band) and the same routing paths and multipaths in Scenario 1 were used. The average end-to-end transmission rate for all paths and multipaths is reported. The results, shown in Figure 2, indicate an almost linear improvement is gained by adding additional channels to the network in terms of additional throughput. The slopes of the lines (as found by a linear regression through each point set) were 8.0, 8.9 and

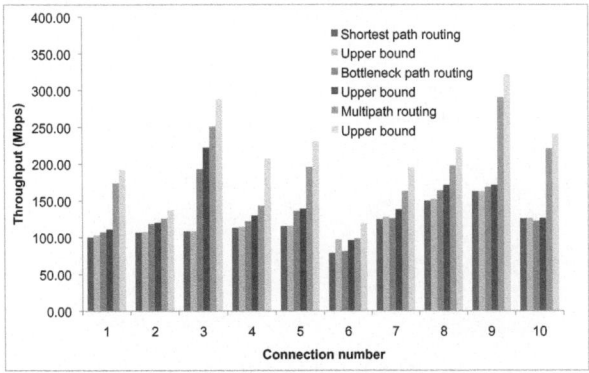

Fig. 1. Schedule performance results for each connection request

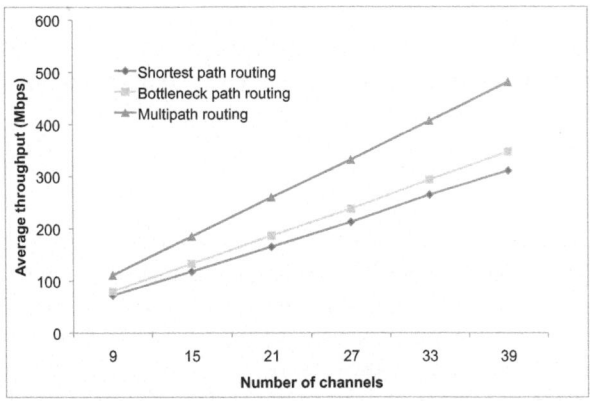

Fig. 2. Average path transmission rate versus the number of available channels

12.3 Mbps / channel respectively for shortest path routing, bottleneck path routing and multipath routing. A possible explanation for the higher slope of multipath routing is that the multipath can make more use of additional channels since the data flow through the multipath is physically more spread out.

6.3 Scenario 3: Network Saturation

In this scenario, we considered how quickly the network would saturate as additional traffic was added to it. We used the original 10 connection pairs from Scenario 1 and also created 10 new random source-destination pairs in the network. The number of channels available in the network was set to 15 (5 from each frequency band). We further assumed that once a connection was established that it would stay active for the remainder of the simulation. The results are shown in Figure 3. As would be expected, multipath routing appears to saturate the network the most quickly, although it is able to achieve the highest

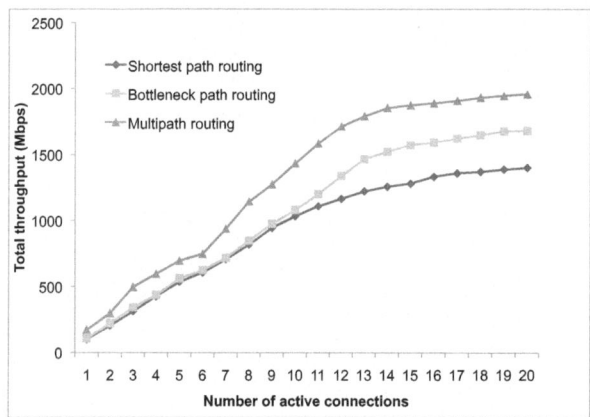

Fig. 3. Network saturation as the number of active connections increases

total throughput. The network saturates the most slowly if shortest path routing is used. Shortest paths in general will use fewer links than the other methods and so will tend to tie up fewer resources per connection.

7   Conclusions

In this paper, we presented a novel polynomial-time constant-factor approximation algorithm for the problem of scheduling transmissions along a multipath in a cognitive radio mesh network and a heuristic routing algorithm for finding multipaths. According to the simulation results, the end-to-end throughput provided by our scheduling algorithm is always close to a computed upper bound on the optimal solution. Our proposed bottleneck path routing approach has improved performance over shortest paths and the proposed multipath routing algorithm can achieve almost twice the performance of shortest path routing. The results demonstrate the potential of using cognitive radios to share spectrum on a non-interfering basis with primary users, while at the same time offering substantial throughput to secondary users.

Acknowledgements. This research was supported in part by a Montana State MBRCT grant 09-23 and NSF grants CNS-0845776, CNS-0721880 and CNS-0624874.

References

1. Akyildiz, I.F., Wang, X., Wang, W.: Wireless mesh networks: a survey. Journal of Computer Networks 47(4), 445–487 (2005)
2. Akyildiz, I.F., Lee, W.-Y., Vuran, M.C., Mohanty, S.: NeXt generation/dynamic spectrum access/cognitive radio wireless networks: a survey. Journal of Computer Networks 50(13), 2127–2159 (2007)
3. Bartal, Y., Byers, J., Raz, D.: Fast, Distributed Approximation Algorithms for Positive Linear Programming with Applications to Flow Control. SIAM Journal of Computing 33(6), 1261–1279 (2004)

4. Brik, V., Rozner, E., Banarjee, S., Bahl, P.: DSAP: a protocol for coordinated spectrum access. In: Proceedings of IEEE DySPAN 2005, pp. 611–614 (2005)
5. Cao, L., Zheng, H.: Distributed spectrum allocation via local bargaining. In: Proceedings of IEEE SECON 2005, pp. 475–486 (2005)
6. Hou, Y.T., Shi, Y., Sherali, H.D.: Optimal spectrum sharing for multi-hop software defined radio networks. In: Proceedings of IEEE Infocom 2007, pp. 1–9 (2007)
7. Huang, J., Berry, R.A., Honig, M.L.: Spectrum sharing with distributed interference compensation. In: Proceedings of IEEE DySPAN 2005, pp. 88–93 (2005)
8. Khalife, H., Ahuja, S., Malouch, N., Krunz, M.: Probabilistic path selection in opportunistic cognitive radio networks. In: Proceedings of IEEE Globecomm 2008, pp. 1–5 (2008)
9. Karnik, A., Iyer, A., Rosenberg, C.: Throughput-optimal Configuration of Fixed Wireless Networks. IEEE/ACM Transactions on Networking 16(5), 1161–1174 (2008)
10. Kompella, S., Wieselthier, J.E., Ephremides, A., Sherali, H.D., Nguyen, G.D.: On optimal SINR-based scheduling in multihop wireless networks. IEEE/ACM Transactions on Networking 18(6), 1713–1724 (2010)
11. Luo, J., Rosenberg, C., Girard, A.: Engineering Wireless Mesh Networks: Joint Scheduling, Routing, Power Control and Rate Adaptation. IEEE/ACM Transactions on Networking 18(5), 1387–1400 (2010)
12. Mumey, B., Zhao, X., Tang, J., Wolff, R.: Transmission Scheduling for Routing Paths in Cognitive Radio Mesh Networks. In: Proceedings of IEEE SECON 2010, pp. 1–8 (2010)
13. Olariu, S.: An optimal greedy heuristic to color interval graphs. Inf. Process. Lett. 37(1), 21–25 (1991)
14. Schneider, J., Wattenhofer, R.: A new technique for distributed symmetry breaking. In: Proceeding of ACM PODC 2010, pp. 257–266 (2010)
15. Shi, Y., Hou, Y.T.: A distributed optimization algorithm for multi-hop cognitive radio networks. In: Proceedings of IEEE Infocom 2008, pp. 1292–1300 (2008)
16. Tang, J., Misra, S., Xue, G.: Joint spectrum allocation and scheduling for fair spectrum sharing in cognitive radio wireless networks. Journal of Computer Networks 52(11), 2148–2158 (2008)
17. Wan, P.-J.: Multiflows in multihop wireless networks. In: Proceedings of MobiHoc 2009, pp. 85–94 (2009)
18. Wang, F., Krunz, M., Cui, S.: Spectrum sharing in cognitive radio networks. In: Proceedings of IEEE Infocom 2008, pp. 1885–1893 (2008)
19. Xi, Y., Yeh, E.M.: Distributed algorithms for spectrum allocation, power control, routing, and congestion control in wireless networks. In: Proceedings of ACM MobiHoc 2007, pp. 180–189 (2007)
20. Xin, C., Xie, B., Shen, C.-C.: A novel layered graph model for topology formation and routing in dynamic spectrum access networks. In: Proceedings of IEEE DySPAN 2005, pp. 308–317 (2005)
21. Yang, Y., Kravets, R.: Contention-aware admission control for ad hoc networks. IEEE Transactions on Mobile Computing 4(4), 363–377 (2005)
22. Yuan, Y., Bahl, P., Chandra, R., Moscibroda, T., Wu, Y.: Allocating dynamic time-spectrum blocks in cognitive radio networks. In: Proceedings of ACM MobiHoc 2007, pp. 130–139 (2007)
23. Zhao, Q., Tong, L., Swami, A.: Decentralized cognitive MAC for dynamic spectrum access. In: Proceedings of IEEE DySPAN 2005, pp. 224–232 (2005)
24. Zheng, H., Peng, C.: Collaboration and fairness in opportunistic spectrum access. In: Proceedings of IEEE ICC 2005, pp. 3132–3136 (2005)

On the Use of Cooperation as an Energy-Saving Incentive in Ad Hoc Wireless Networks

Maurizio D'Arienzo[1], Sabato Manfredi[2], Francesco Oliviero[2],
and Simon Pietro Romano[2]

[1] Dipartimento di Studi Europei e Mediterranei, Seconda Università di Napoli, Italy
maudarie@unina.it
[2] University of Napoli Federico II, Napoli, Italy
{sabato.manfredi,folivier,spromano}@unina.it

Abstract. In this paper we show how cooperation can improve the overall energy efficiency of an ad hoc network. By exploiting a behavior-tracking algorithm inspired by the results of game theory and allowing traffic to be forwarded only towards cooperative nodes, we show how we can dramatically reduce power wastage at the same time maximizing goodput. Under this perspective, cooperation can definitely be seen as an incentive for all nodes, since it allows to optimize one of the most crucial parameters impacting the performance of ad hoc networks.

1 Introduction

Ad hoc networks are composed of several nodes with wireless connection capability. Differently from wired networks, in an ad hoc environment each node is an end system and a router at the same time. A transmission between a sender and a receiver happens with the help of one or more intermediate nodes that are requested to relay packets according to routing protocols designed for this kind of networks. A blind trust agreement among nodes makes it possible the right message forwarding. Actually, a generic node of the network should be able to decide whether or not to trust the other nodes. This obviously calls for a capability of each single node to somehow interpret (or, even better, predict) the behavior of the other nodes, since they represent fundamental *allies* in the data transmission process. The situations in which a decision of a part depends on the predicted behavior of another part have been elegantly studied in game theory. Game theory has been already applied [3] [4] [5] to ad hoc networks with interesting results. The basic assumption is that all the players follow a rational behavior and try to maximize their payoff. The simplest games see the involvement of only two players who have to decide whether to cooperate or defect with the others. The best solution may not maximize the payoff, but can reach an *equilibrium* as proposed by Nash. One of the versions of this game is known as *prisoner's dilemma* and has an equilibrium in case both users decide to defect. This is true for the game played only one time while in its iterated version the situation is more complex and even cooperation can be convenient. In case of ad an hoc network, the player is a node that needs to cooperate with the others to send its traffic. However some nodes can decide to defect for a number of unspecified reasons and, as a first need, the other nodes should be informed of their behavior in order to react in the most appropriate way.

In this paper, we show how cooperation can be perceived by nodes as an incentive, thanks to the fact that it helps save the overall amount of energy needed for data transmissions. Differently from recent works proposed in the literature, which aim at making the routing process become *natively* aware of the energy-related parameters, we herein propose a different approach, by leveraging cooperation in order to improve the overall energy efficiency of an ad hoc network without modifying the existing routing protocol. We present in the paper an algorithm to identify and isolate defecting nodes. The algorithm takes inspiration from the results of game theory and keeps a local trace of the behavior of the other nodes. In case defecting nodes are identified, different countermeasures (i.e. not relaying packets coming from defecting nodes) can be adopted.

The paper is organized in six Sections. Section 2 deals with both background information and related work. Section 3 presents the algorithm we designed to infer behavioral information about the network nodes, whose implementation is described in Section 4. Results of the experimental simulations we carried out are presented in Section 5, while Section 6 provides concluding remarks and proposes some directions of future work.

2   Background and Related Work

In this section we try to shed light on the context of our contribution, by properly defining the scope of our research. We focus on the most important aspect of our contribution, namely cooperation. Indeed, as we already pointed out, cooperation is a fundamental subject of our recent research and is herein studied under one of its most challenging facets, i.e. its use as an incentive for all the nodes of the network, thanks to the significant performance improvements that it entails in terms of energy savings associated with data transmissions.

Cooperation of nodes involved in an ad hoc network is usually induced because the efforts related to the offered services are compensated with the possibility to request a service from the other nodes. However current ad hoc network protocols do not provide users with guarantees about the correct behavior of other nodes that can potentially decide to act as parasites. Several works have identified the problem of stimulating cooperation and motivating nodes towards a common benefit. The main solutions rely on a virtual currency or on a reputation system, and more recently on game theory.

Virtual currency systems [6] [7] give well behaving users a reward every time they regularly relay a packet. They can then reuse the reward for their transmissions as long as they have a credit. The first issue of such systems is related to the need of a centralized server to store all the transactions among the users. Reputation systems repeatedly monitor and build a map of trustworthy nodes on the basis of their behavior [1] [2] [5] [8]. These systems distinguish between the *reputation*, which rates how well a node behaved, and *trust*, which represents how honest a node is. Most of these systems consider the reputation value as a metric of trust . A node is refrained from relaying a packet coming from untrusted nodes, which are then excluded from the network operations. Several issues are related to the use of these systems. First, each node needs to maintain a global view of the reputation values, with considerable caching. Some proposals keep local information, others disseminate reputations to other nodes, with an increased overhead due to the exchange of such messages. Reputation values can be modified, forged or lost during operations, and they can differ from node to node, which can bring to inconsistencies.

To overcome some of these issues, it has been proposed to model the nodes taking part to an ad hoc network with game theory, a branch of applied mathematics which witnessed a great success thanks to the application of its results to a wide selection of fields, including social sciences, biology, engineering and economics. Game theory covers different situations of conflicts regarding, in a first attempt, two agents (or *players*), and in the generalized version, a population of players. Each of these players expects to receive a reward, usually named *payoff*, at the end of the game. The basic assumption is that all the players are self interested and rational: given a utility function with the complete vector of payoffs associated with all possible combinations, a rational player is always able to place these values in order of preference even in case they are not numerically comparable (e.g. an amount of money and an air ticket). This not necessarily means that the best value will be selected, since the final reward of each player is strongly dependent on the decision of the other players. Each player is then pushed to plan a *strategy*, that is a set of actions aiming at a total *payoff* maximization, provided that he is aware that the other players will try to do the same. Games are now classified according to various properties. Here we are mainly interested in the difference between *cooperative* and *non-cooperative* games as well as the difference between *strategic games* (played once) and *extensive games* (played many times).

One of the fundamental problems of game theory is known as *prisoner's dilemma*, which can be represented in the matrix format of Fig. 1: two suspects of a crime are arrested and jailed in different cells with no chance to communicate between each other. They are questioned by the police and receive the same deal: if one confesses (*defect*) and the other stays silent (*cooperate*), the first is released, the second is convicted and goes to prison with a sentence of 10 years, the worst; if both stay silent (*cooperate*), they go to prison for only 1 year; if both testify against the other (*defect*) they go to prison with a sentence of 5 years. The situation in which they both stay silent (*cooperate*) is the more convenient to both of them; however, it was demonstrated that a rational behavior is to confess (*defect*) and receive the sentence of 5 years, and this situation represents the only equilibrium, as first introduced by Nash [14] [13]. Hence, the prisoner's dilemma falls in the field of strategic non-cooperative games. In its basic form the prisoner's dilemma is played only once and has been applied to many real life situations of conflict, even comprising thorny issues of state diplomacy.

Another version of the prisoner's dilemma is played repeatedly rather than just once and is known as iterated prisoner's dilemma (ITD), which turned out to be a cooperative game under certain circumstances [9][10]. The goal of both players still is the maximization of their payoff, as the cumulated payoff earned at each stage. If the number of rounds is finite and known in advance, the strategy of always defecting is still the only situation of equilibrium and the game is still non-cooperative. However, in case the number of repetitions is infinite, it was demonstrated that the choice to always defect is not the only equilibrium as even the choice of cooperating may be an equilibrium. In this case, one of the strategies that let players maximize their payoff is the so-called *Tit for Tat* game, in which each player repeats the past behavior of the other player: a player is keen to cooperate if the other node behaved correctly the last time, otherwise it defects. If we consider the first five tournaments of a two players game, a player who defects (D) against a cooperative (C) player adopting the tit for tat strategy would play (D,D,D,D,D) and earn $(0, -5, -5, -5, -5) = -20$. If the first player decides to cooperate two times out of five (D,D,C,C,D), he would earn $(0, -5, -10, -1, 0) = -16$. In case he always cooperates, his payoff would be $(-1, -1, -1, -1, -1) = -5$, which is the best he can achieve. So, continued cooperation for the iterated prisoner's dilemma also yields the best payoff. Despite this benefit, the main result of the tit for tat strategy is that it stimulates the cooperation. We base our algorithm to mitigate the node selfishness on the results of this version of the game.

Game theory has already been applied to the study of ad hoc networks. One of the first proofs of the improvements produced by cooperation in such networks is presented in [3]. The authors first introduce a normalized acceptance rate (NAR) as the ratio between the successful relays provided to the others and the relay requests made by the node. Then they propose two models, namely GTFT (Generous Tit for Tat) and m-GTFT for the case of multiple players, to give the (rational) nodes the chance to make a decision concerning the possibility to cooperate or defect with other nodes, and they analytically demonstrate that these models represent a Nash equilibrium. In such a situation, a node does not improve its NAR to the detriment of the others. Also, at the opposite of reputation schemes, each node can maintain per session rather than per packet information, thus leading to a scalable solution.

In [15] the authors prove the selfishness property of the nodes in a MANET by using the Nash equilibrium theorem [13]. They define a generic model for node behavior which takes into account also energy consumption due to the transmission process. By adopting a punishment based technique they prove that it is possible to escape from the theoretically unique equilibrium point of non-cooperation and to enforce a cooperation strategy under specific conditions.

In [16] the authors also focus on forwarding mechanisms. They provide a model for node behavior based on game theory in order to determine under which conditions cooperation with no incentives exists. They prove that network topology and communication patterns might significantly help enforce cooperation among nodes.

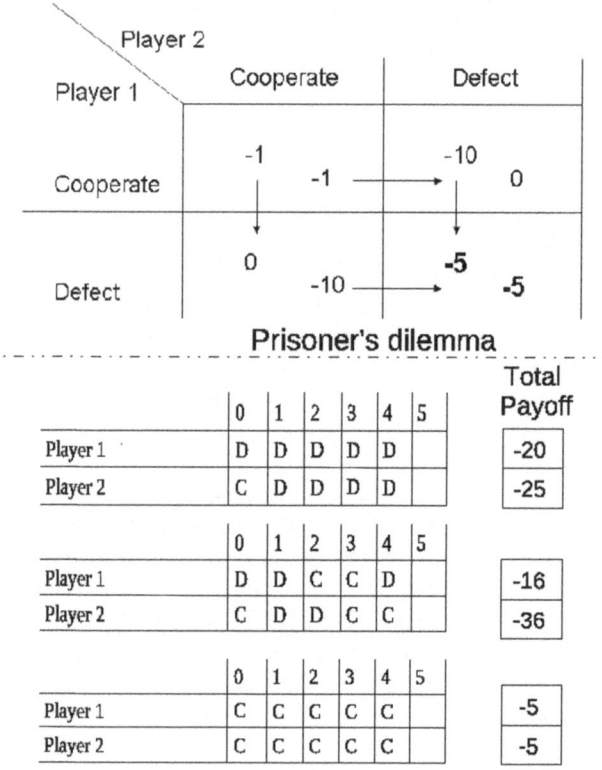

Fig. 1. The tit-for-tat strategy in action

Game theory has been also used to improve routing algorithms in wireless networks. An actual implementation of a game theory model in the AODV routing protocol with two distinct approaches has been proposed in [4]. The first plays a deterministic tit for tat game and the second a randomized version of the same game deployed with a genetic algorithm. In both cases, they achieve better performance in terms of experienced delay and packet delivery ratio in case of cooperation of nodes. The models are tested in a simulated environment and rely on static distribution of nodes' behavior profiles while not supporting a mechanism for a dynamic adaptation to changed situations.

3 A Novel Behavior-Tracking Approach

In an ad hoc network, the number of nodes and links can change over time, so we consider the number of nodes $N(t)$ as a function of time t. We also define a dynamic array $C(t)$ of $N(t)$ elements for each node of the network. The generic element $c_i(t)$ of $C(t)$ assumes the values (UNKNOWN, COOPERATE, DEFECT) meaning that the behavior of node i at time t is respectively unknown, cooperative or non cooperative. At time $t = 0$ all the values are set to UNKNOWN, since at the beginning each node is not aware of the behavior of the other nodes. We herein propose to introduce a routing control strategy at each node based on the game theory framework described earlier. Basically, the control algorithm firstly identifies the cooperative nodes (Detection Phase) and then reacts in the most appropriate way in order to give priority to packets generated by cooperative nodes. Specifically the control algorithm is composed of a *Detection Phase*, followed by a *Reaction Phase*. The former works as follows. Suppose the generic node s of the network needs to send some traffic to the destination d. The first task is to discover an available path, if it exists, to reach the destination. To this purpose, we consider a source based routing protocol capable of discovering a list $A(t)_{(s,d)i}$, $\forall i : 0 < i < P$, of P multiple paths. All the nodes in the list $A(t)_{(s,d)i}$ are considered under observation and marked as probably defecting in the array $C(t)$ unless a positive feedback is received before a timeout expires. The sender s starts sending his traffic along all the discovered paths. If the destination node generates D acknowledgement messages containing the list of all the nodes $L_{(s,d)i}$ (with $0 < i < D$) traversed, as it happens in some source based routing protocols, the sender s is informed about the behavior of intermediate nodes. For each acknowledgement message received, the sender s can make a final update of the array $C(t)$ by setting the matching elements $c_i(t)$ contained in the list $L_{(s,d)i}$ as cooperative. Notice that the last update overwrites the previous stored values and represents the most recent information concerning the behavior of a node.

Once done with the detection phase, each node is aware of the behavior of other nodes and can enter the reaction phase in the most appropriate way. For example, a node can refuse to relay packets of defecting nodes, or operate a selective operation like queuing their packets and serving them only if idle and not busy with the service requested by cooperative nodes. In this first proposal, we rely on the harsh policy of packet discarding, and this brings to the isolation of defecting nodes. However, a defecting node can even gain trust of other nodes if it starts to cooperate. The array $C(t)$ is not static over time and its values are continuously updated. In fact, due to the dynamic situation of ad hoc networks, the search of available paths is frequently repeated, and the list $A_{(s,d)}$ consequently updated. Hence, if a defecting node decides to cooperate, its identification address will be included in one of the acknowledgement messages $L_{(s,d)i}$ sent to the sender s and its aim to cooperate will be stored in the array $C(t)$.

The situation described here for the pair (s, d) is replicated for all possible pairs of nodes that try to interact, but each node stores only one array $C(t)$ that is updated upon reception of any acknowledgement message, wherever it comes from. Furthermore, not all the packets relayed are checked in order to verify the nodes' behaviors, but only a sample of them, thus keeping the total overhead under control.

4   Algorithm Implementation

The algorithm introduced in the previous section has been implemented in AH-CPN (Ad Hoc Cognitive Packet Network) [11], an existing source based routing protocol for ad hoc networks. AH-CPN is the wireless version of CPN (Cognitive Packet Network) [12], a proposal for a self aware network architecture with native support for QoS.

There are four different kinds of packets in AH-CPN: Smart Packets (SP), Smart Acknowledgements (SA), Dumb Packets (DP), and Dumb Acknowledgements (DA). SPs are lightweight packets sent by a sender towards a destination to discover new paths according to specific QoS goals, e.g. discovering paths that minimize the delay or maximize the throughput. Once at the destination, a SA is generated and sent backwards along the reverse path received in the SP. Finally, the actual data can be sent across the network in a DP, which is prepared with the whole path copied in the DP header. Once the DP reaches its destination, a DA is sent along the reverse path. Notice that differently from IP networks, in CPN the acknowledgements are generated upon reception of each single packet, whatever the transport protocol is. This feature is helpful in the deployment of our algorithm to identify defecting nodes, as we will soon explain. We first modified this protocol to support the search of multiple paths, and then included the new algorithm for the identification of non cooperative nodes.

The basic AH-CPN version looks for one available path, the best in terms of the requested QoS goal. We modified this protocol to search for multiple paths. To this purpose, SPs are initially sent via flooding to collect all the available paths. To prevent loops, SPs are marked with an identification number ID, and those with the same ID touching a node for the second time are discarded. SPs reaching the same destinations with different contents for what concerns the routing map are considered valid, and SAs are sent backward to inform the sender. The sender collects the different SAs and updates its routing table. DPs are sent on a round robin basis. Once the available paths are discovered, the transmission of SPs is not terminated; it is rather repeated periodically for path maintenance, to check if the topology has changed, and in our case also to verify if there is a different configuration concerning the behavior of nodes.

We then provided the multipath source based routing protocol with the support for identification and isolation of defecting nodes. The array $C(t)$ is added and stored in each node and its dimension can change according to the number of nodes active in the ad hoc area. When node a needs to send traffic to node b, SPs are immediately sent in flooding. We make the assumption that non cooperative nodes try to cheat by forwarding inexpensive SPs, that do not carry

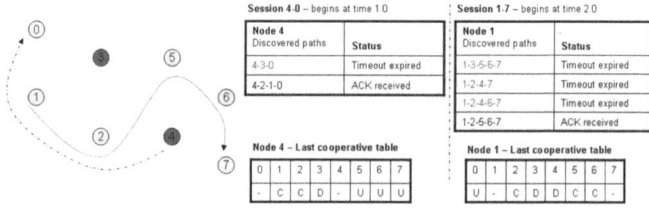

Fig. 2. The simulated testbed

any payload, while they do not relay DPs containing the real data. In case the non cooperative nodes decide to block the SPs forwarding, they are immediately discovered as non cooperative and have no chance to cheat. In this scenario, every time a SP traverses a node, its cognitive map is extended with the label of the visited node. Once at the destination, the complete cognitive map is copied into the DA and sent back to the sender along the reverse path. Obviously, this is repeated for all the discovered paths, so at the end of the process node a has a complete knowledge of all the available paths, also those comprising cheating nodes, and these are all stored in $A(t)_{(a,b)}$. At the time of the first transmission, the real data are packed in multiple DPs and sent along all the available paths on a round robin basis, but the interested cheating nodes will not rely ay them. Since in CPN a destination b must send an acknowledgement message DA whatever the transport protocol is, node a will receive only the DAs containing the successful paths, i.e. those without cheating nodes. This information, as described before, helps finalize the array $C(t)$ with the list of cooperative and defecting nodes, and the traffic is sent only along the path or the paths composed of cooperative nodes rather than towards all the available paths. When one of the cheating nodes requests the relaying of a message to node a, it is aware of his past behavior and can decide to drop all its packets, while it can regularly relay packets coming from cooperative users.

The situation concerning the cooperation and the selection of paths is not static and can change over time, so isolated nodes are not banned forever from the network. Although the traffic from a node is delivered only along paths composed of cooperative nodes, sending nodes continue to check periodically the paths containing the defecting nodes. Should a defecting node decide to change its behavior and begin to cooperate, the routing protocol soon detects this change and admits again the node to the transmission of flows. This way, a node reacts following a *Tit for Tat* strategy.

5   Experimental Results

The introduction of a system able to detect defecting nodes composing a wireless ad hoc network can lead to a better distribution of the energy consumed by each node. To show this effect, we tested the proposed system with the ns-2 simulator. The current implementation of the algorithm relies on a dedicated multiple path ad hoc routing protocol that supports the explicit acknowledgement of packets regularly received at the final destinations. To highlight the robustness of the algorithm, we designed a scenario associated with several working conditions, on a simple wireless testbed composed of 8 nodes (see Fig.2), labeled from 0 to 7. In such network we set up the following conditions: (i) node 3 defects all the time; (ii) the behavior of node 4 dynamically changes over time; (iii) all the other nodes are cooperative. The duration of the experiments is set to 12 minutes. The defection of a node means that the relay of traffic to serve other nodes is totally stopped, so the percentage of node 3's cooperation is 0% (of the total time). As far as node 4 is concerned, five situations are considered, most of them offering the other nodes the chance to reply with a *tit for tat* strategy: (i) node 4 never cooperates, hence requests of relay are never forwarded and the percentage of cooperation is 0%; (ii) node 4 follows a switching behavior: each 3 minutes interval, it defects for the first 2 minutes and then cooperates for the remaining minute, for a total percentage of cooperation of 33%; (iii) node 4 still switches its behavior: each 2 minutes interval, it defects and cooperates in equal parts, arriving at a cooperation percentage of 50%; (iv) node 4 switches its behavior in a way that is opposite to the one described in the second item of this list: each 3 minutes interval, node 4 defects for the first minute and cooperates for the last 2 minutes, hence cooperating for 75% of the time; (v) node 4 always cooperates: all relay requests are served (for a final percentage of cooperation of 100%).

Two equal sessions of constant bit rate traffic are activated between node 4 and node 0 and node 1 and node 7, respectively at time 1.0 and at time 2.0. In the ideal situation of all cooperating nodes, the shortest paths would be $(4, 3, 0)$ and $(1, 2, 4, 7)$. However, node 3 is always defecting, so the path $(4, 3, 0)$ turns out to be unavailable and the traffic coming from node 4 is forced along the other available path $(4, 2, 1, 0)$. As long as node 1 does not generate traffic, it does not have the chance to track the behavior of node 4, so the relay requests coming from node 4 are regularly served. At time 2.0 node 1 begins the discovery of paths to reach node 7. Besides the other choices, the best path $(1, 2, 4, 7)$ is soon discovered and selected to immediately generate traffic. If node 4 follows a switching behavior, then node 1 has the chance to react in compliance with the *tit for tat* strategy. Notice that in case node 4 is in a defecting state, node 1 still can send traffic to the destination along the path $(1, 2, 5, 6, 7)$.

In the first graph we evaluate the goodput of node i as the ratio $G_i(t) = r_i(t)/s_i(t)$ at the end of the experiment ($t = 12min$) between the number of bytes correctly received at destination and the total number of bytes sent. The x axis represents the percentage of node 4's cooperation, the y axis is the final goodput $G_i(t)$. On the left hand side of Fig. 3, node 4 is fully defecting; the same applies to node 3. Traffic from node 4 towards node 0 is regularly sent between time 1.0 and time 2.0 because node 1 did not generate any request and did not yet check the behavior of the other nodes. At time 2.0, however, node 1 tries to send traffic to node 7 and hence has the chance to verify the behavior of the other nodes. Among the other discovered paths, it realizes that paths comprising nodes 4 and 3 are not working, so as soon as the timeout expires it marks nodes 3 and 4 as defecting and immediately stops relaying traffic coming from node 4. The final goodput G_1 of node 1 is closer to the ideal value because the alternative path $(1, 2, 5, 6, 7)$ is soon discovered and used for the entire duration of the experiment. Goodput G_4 is instead severely reduced. As node 4's percentage of cooperation increases up to 100%, goodput G_4 increases until it reaches a value close to goodput G_1 when there is full cooperation. Although node 3's defection makes the path $(1, 3, 0)$ unavailable, the routing protocol discovers the alternative path $(4, 2, 1, 0)$ composed of cooperative nodes, while the shortest $(1, 2, 4, 7)$ is regularly available in this case. This is the only situation in which node 4 maximizes its goodput. In the intermediate cases the trend is linear

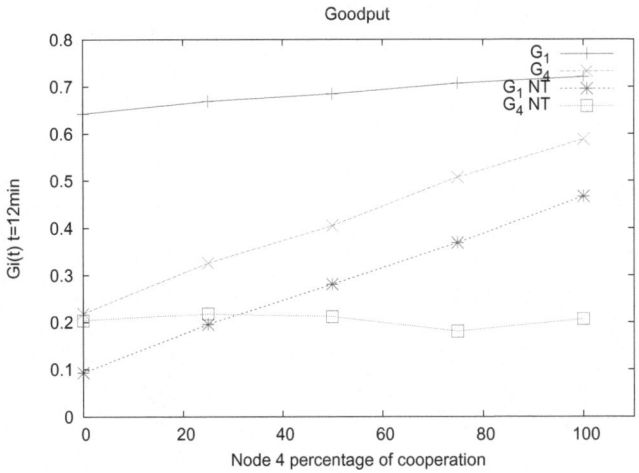

Fig. 3. Goodputs of Node 1 and Node 4

and clearly demonstrates the correct implementation of the *tit for tat* reaction mechanism, as node 1 cooperates only when node 4 does the same. Goodput G_1 remains more or less unaltered independently of node 4's behavior, thanks to the fact that node 1 has a chance to discover alternative cooperative paths. We compared these results with the situation in which the nodes are unable to detect the defecting behavior. We mark these sessions with *NT* in the same figure 3. The situation is now opposite to the previously analyzed case because goodput G_4 outperforms G_1 in the case of node 4's full defection. Node 1 is now unaware of node 4's defection; hence, while its traffic is not relayed, it regularly relays the incoming packets having node 4 as source. Anyway, both goodputs G_1 and G_4 are lower than in the previous case. This time the lack of tracing of nodes defection affects even node 4's performance, because such node tries to forward traffic not only along the path $(4, 2, 1, 0)$ but also along the uncooperative path $(4, 3, 0)$, which explains the halved final goodput.

We now introduce a new function to evaluate node energy consumption with respect to the various levels of cooperation. We define the parameter S_i computed at the end of the experiments as:

$$S_i = \frac{Ec_i}{(s_i + rl_i)} * \frac{s_i}{r_i}$$

being Ec_i the energy consumed by node i, s_i the total number of bytes sent to the destination, rl_i the number of bytes relayed from node i, and r_i the bytes correctly received at destination. S_i has a dimension of $[Joule/bytes]$ and represents the energy spent by a node to successfully deliver a byte to the destination. We show in figure 4 the value S_1 and S_4 calculated at the end of the experiments and for all the aforementioned combinations of cooperation. From the figure we can observe a significant difference between the energy consumed to deliver one

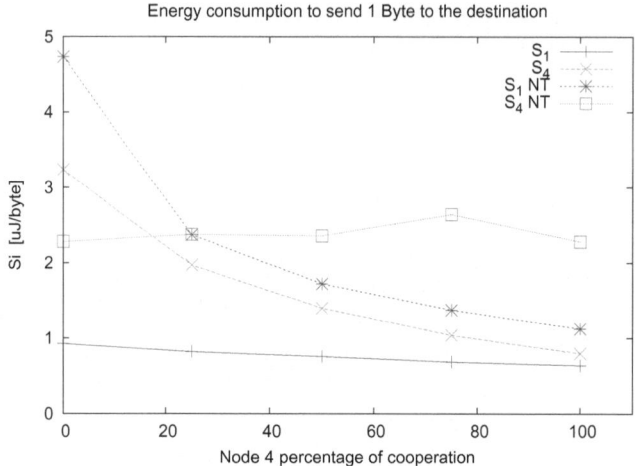

Fig. 4. Energy consumed to deliver a single byte

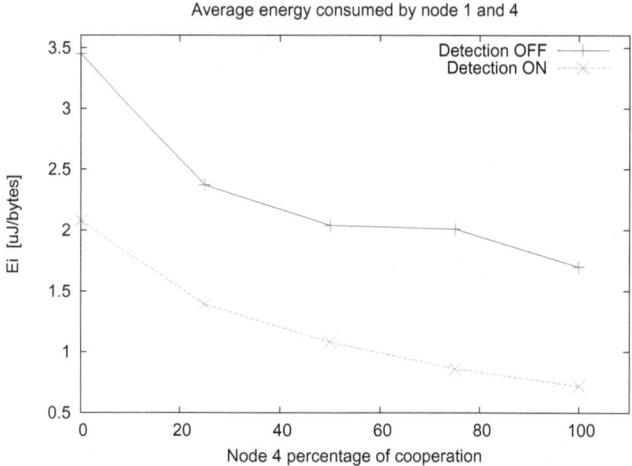

Fig. 5. Average consumed energy of node 1 and 4

byte to the destination in favor of node 1 when node 4 does not cooperate at all. This difference reduces as node 4's percentage of cooperation increases, and it becomes negligible when both cooperate. Even in this case, the trend along the intermediate situations seems linear.

We compared again these results with the analogue experiments executed without the detection of defecting nodes. Energy consumption at node 1 is severely affected, especially when node 4 does not cooperate at all. The convenience of node 4 limits to the single case of total defection at the expenses of the unaware node 1; however, starting from the next experiment and until full cooperation is reached, node 1 again saves more energy than node 4.

Finally, notice that if we consider the average energy consumed by both node 1 and node 4 in case the detection of defecting nodes is enabled, such value is always lower compared to the case when detection is disabled, as showed in Fig. 5. The energy consumed by node 1, the only node which always cooperates, is always lower than that consumed by node 4. Furthermore, node 4's consumption reaches its lowest value when the cooperation is full.

6   Conclusions

In this paper we demonstrated through simulations that cooperation actually acts as an incentive for ad hoc network nodes, since it allows for a lower average energy expenditure per byte transmitted. We also studied the positive impact of cooperation on goodput, which is considered a key performance indicator for any networked environment. Namely, we proved that the resulting network equilibrium achieved in the presence of cooperative nodes increases fairness in terms of energy consumed per unit of successfully delivered packets.

This work is clearly just a first attempt at studying the many facets of cooperation in ad hoc networks. Among the numerous improvements that we identified and which represent directions of our future work, we firstly mention a more detailed analysis of the dependence of the performance improvements deriving from cooperation on the specific network topology taken into account. Apart from this, we also intend to study how the specific location of a node in the ad hoc network topology affects its performance and consequently its willingness to cooperate. This requires that a thorough analysis of the tradeoff between relaying other nodes' packets and sending one's own data is conducted.

References

1. Olivero, F., Romano, S.P.: A reputation-based metric for secure routing in wireless mesh networks. In: IEEE GLOBECOM 2008, pp. 1–5 (December 2008)
2. Mandalas, K., Flitzanis, D., Marias, G.F., Georgiadis, P.: A survey of several cooperation enforcement schemes for MANETs. In: IEEE Int. Symp. on DOI, pp. 166–171 (2005)
3. Srinivasan, V., Nuggehalli, P., Chiasserini, C.F., Rao, R.R.: Cooperation in wireless ad hoc networks. In: INFOCOM 2003, vol. 2, pp. 808–817 (April 2003)
4. Komathy, K., Narayanasamy, P.: Trust-based evolutionary game model assisting aodv routing against selfishness. J. Netw. Comput. Appl. 31(4), 446–471 (2008)
5. Marti, S., Giuli, T.J., Lai, K., Baker, M.: Mitigating routing misbehavior in mobile ad hoc networks. In: ACM MobiCom 2000, New York, USA, pp. 255–265 (2000)
6. Zhong, S., Yang, Y., Chen, J.: Sprite: A simple, cheat-proof, credit-based system for mobile ad hoc networks. In: INFOCOM 2003, vol. 3, pp. 1987–1997 (2003)
7. Buttyán, L., Hubaux, J.P.: Stimulating cooperation in self-organizing mobile ad hoc networks. Mob. Netw. Appl. 8(5), 579–592 (2003)
8. Buchegger, S., Le Boudec, J.-Y.: Performance analysis of the confidant protocol. In: ACM MobiHoc 2002, New York, USA, pp. 226–236 (2002)
9. Axelrod, R.: The Evolution of Cooperation. Basic Books (1988)

10. Axelrod, R., Dion, D.: The further evolution of cooperation. Science 242(4884), 1385–1390 (1988)
11. Gelenbe, E., Lent, R.: Power-aware ad hoc cognitive packet networks. Ad Hoc Networks 2(3), 205–216 (2004)
12. Gelenbe, E., Lent, R., Xu, Z.: Design and performance of cognitive packet networks. Perform. Eval. 46(2-3), 155–176 (2001)
13. Nash, J.: Non-Cooperative Games. The Annals of Mathematics 54(2), 286–295 (1951)
14. Nash, J.F.: Equilibrium Points in n-Person Games. Proceedings of the National Academy of Sciences of the United States fo America 36(1), 48–49 (1950)
15. Urpi, A., Bonuccelli, M., Giordano, S.: Modelling Cooperation in Mobile Ad Hoc Networks: A Formal Description of Selfishness. In: Proceedigns of Modeling and Optimization in Mobile, Ad Hoc and Wireless Networks (2003)
16. Félegyházi, M., Hubaux, J.P., Buttyán, L.: Nash Equilibria of Packet Forwarding Strategies in Wireless Ad Hoc Networks. IEEE Transaction on Mobile Computing 5(5), 463–476 (2006)

Cooperation Policy Selection
for Energy-Constrained Ad Hoc Networks
Using Correlated Equilibrium

Dan Wu[1], Jianchao Zheng[1], Yueming Cai[1,2], Limin Yang[1], and Weiwei Yang[1]

[1] Institute of Communications Engineering, PLA University of Science and
Technology, Nanjing, China
[2] National Mobile Communications Research Laboratory, Southeast University,
Nanjing, China

Abstract. Energy efficiency is crucial for energy-constrained ad hoc
networks. Cooperative communication can be applied to significantly reduce energy consumption. Due to the selfishness and the self-organization
of nodes, the relay requests can not always be accepted by potential relay nodes with only local information, and the network overall performance can not always be improved in a distributed way. In this work, we
present a distributed cooperation policy selection scheme which allows
nodes to autonomously make their own cooperation decisions to achieve
the global max-min fairness in terms of energy efficiency. Specifically,
since the correlated equilibrium can achieve better performance by help-
ing the noncooperative players coordinate their strategies, we model a
correlated equilibrium-based cooperation policy selection game, where
the individual utility function is designed from the global energy efficiency perspective. We derive the condition under which the correlated
equilibrium is Pareto optimal, and propose a distributed algorithm based
on the regret matching procedure that converges to the correlated equilibrium. Simulation results are provided to demonstrate the effectiveness
of the proposed scheme.

Keywords: ad hoc networks, cooperative communication, energy efficiency, outage probability, game theory, correlated equilibrium.

1 Introduction

Energy efficiency is of great importance to energy-constrained ad hoc networks,
and the cooperative transmission technique is now widely considered as a promising approach to achieve energy efficiency [1]. The choice of cooperation policies
is essential to exploit this energy saving potential of cooperation, however, existing methods are almost based on maximizing the throughput [2], minimizing
the symbol error rate (SER) [3], etc. Since the nodes in ad hoc networks are distributed and selfish, how to select proper cooperation policies from a distributed
perspective has been an important issue to be solved. Game theory provides a
highly appealing mathematical tool for addressing the issue of node cooperation

Joel J.P.C. Rodrigues et al.: (Eds.): GreeNets 2011, LNICST 51, pp. 161–170, 2012.
© Institute for Computer Sciences, Social Informatics and Telecommunications Engineering 2012

in ad hoc networks, such as [4,5]. Note that most of these works focus on the concept of Nash equilibrium in specific resource allocation games. However, the Nash equilibrium does not always lead to the best performance for the nodes which are distributed, competitive, and equipped with a low-level awareness of neighboring environments.

In this work, we aim at achieving a global objective (the max-min fairness in terms of energy efficiency with outage performance constraint) using a distributed scheme (cooperation policy selection by individual nodes) in energy-constrained cooperative ad hoc networks. Firstly, we transform a global objective into local objective function according to the relationship between them. Then, the resulting individual objective function triggers a cooperation policy selection game. Particularly, we focus on the correlated equilibrium to analyze the outcome of the proposed game. Since the correlated equilibrium directly considers the ability of nodes to coordinate actions, it is a better solution compared to the non-cooperative Nash equilibrium, and is naturally attractive for distributed adaptive algorithms to solve discrete problems. Recently, several wireless networking problems have been characterized by using the correlated equilibrium concept [6–8]. Furthermore, we prove that the correlated equilibrium of the proposed game is Pareto optimal in some specific cases. Also, we propose an algorithm based on the regret matching procedure to obtain the correlated equilibrium in a distributed manner. The resulting correlated equilibrium can help us select proper cooperation policies.

2 Syetem Model and Problem Formulation

2.1 System Model

We consider an energy-constrained ad hoc network consisting of N nodes, where each node is endowed with a single antenna and a half-duplex transceiver. For a cooperative network model, a node plays the role of the source node (s) to send a number of data to a destination node (d), and the remaining $N-1$ nodes form the set of potential relay nodes, denoted by $\mathcal{R}_p = \{r_j\}$. In general, the communication between the source and destination nodes is divided into two phases, i.e., a local broadcasting transmission and a long-haul cooperative transmission. In terms of both phases, the channels are modeled by a path loss exponent δ and frequency flat Rayleigh fading (i.e., $h_{s,r_j} \sim \mathcal{CN}(0,1)$, $h_{s,d} \sim \mathcal{CN}(0,1)$ and $h_{r_j,d} \sim \mathcal{CN}(0,1)$ are unitary power, Rayleigh fading coefficients).

During the first phase, the source node chooses $n_t - 1$ nodes to form the set of relay nodes $\mathcal{R}$, and broadcasts its data to them. Due to the broadcasting nature of the wireless channel and the goal of guaranteeing the nodes in $\mathcal{R}$ decode correctly, the capacity region of the local broadcasting transmission is constrained to:

$$\min_{r_j \in \mathcal{R}} \left\{ \frac{1}{2} \log_2 \left(1 + \frac{p_s^{co,1}}{\sigma^2} \kappa d_j^{-\delta} \left| h_{s,r_j} \right|^2 \right) \right\} \geq \mathcal{C}_{out}, \tag{1}$$

$$\Rightarrow p_{\mathrm{s}}^{\mathrm{co},1} \geq \frac{\left(2^{2\mathcal{C}_{out}} - 1\right)\sigma^2\kappa^{-1}}{\min_{\mathrm{r}_j \in \mathcal{R}}\left\{d_j^{-\delta}\left|h_{\mathrm{s},\mathrm{r}_j}\right|^2\right\}}, \tag{2}$$

where $\mathcal{C}_{out}$ is the outage capacity, $p_{\mathrm{s}}^{\mathrm{co},1}$ is the power needed for broadcasting, σ^2 is the Gaussian noise variance, d_j is the local distance between the source node and r_j, and κ is a constant which depends on the propagation environment.

During the second phase, the relay nodes form the virtual MISO with the source node based on distributed space time codes (DSTC), and jointly transmit the data to the destination node with the transmission power $p^{\mathrm{co},2}$. Each transmitting member has the same transmission power, i.e., $p_{\mathrm{s}}^{\mathrm{co},2} = p_{\mathrm{r}_j}^{\mathrm{co},2} = p^{\mathrm{co},2}/n_t$, $\forall \mathrm{r}_j \in \mathcal{R}$, $|\mathcal{R}| = n_t - 1$. Furthermore, the outage probability is

$$P_{\mathrm{out}} = \Gamma\left(n_t, \frac{\left(2^{2\mathcal{C}_{out}} - 1\right)\sigma^2\kappa^{-1}d^{\delta}}{p^{\mathrm{co},2}/n_t}\right), \tag{3}$$

where d is the long-haul transmission distance between the transmitting members and the destination node, and $\Gamma\left(n_t, b\right) = \frac{1}{(n_t-1)!}\int_0^b x^{n_t-1}e^{-x}dx$.

2.2   Problem Formulation

In order to determine globally optimal cooperation policies in terms of energy efficiency, it is necessary to deal with two tricky problems: i) the CSI should be acquired for all links and for all time; ii) improving energy efficiency focuses on not only reducing the energy consumption of the whole network, but also making balanced use of each nodes energy. To this end, we denote outage performance as the target, which depends on large scale channel effects and models small scale fading via using its statistical description. Moreover, we should formulate the optimal cooperation policy selection problem from a global and max-min fairness perspective. Hence, we describe the optimization problem as maximizing the residual energy of worst-off node with a outage performance constraint, i.e,

$$\max_{\mathcal{R},\mathbf{p}}\left\{u\left(\mathcal{R},\mathbf{p}\right) = \min\left\{E_{\mathrm{s}} - \frac{1}{2}\left(p_{\mathrm{s}}^{\mathrm{co},1} + p_{\mathrm{s}}^{\mathrm{co},2}\right), E_{\mathrm{r}_j} - \frac{1}{2}p_{\mathrm{r}_j}^{\mathrm{co},2}, \forall \mathrm{r}_j \in \mathcal{R}\right\}\right\}, \tag{4}$$

subject to

$$p_{\mathrm{s}}^{\mathrm{co},1} \geq \frac{\left(2^{2\mathcal{C}_{out}} - 1\right)\sigma^2\kappa^{-1}}{\min_{\mathrm{r}_j \in \mathcal{R}}\left\{d_j^{-\delta}\left|h_{\mathrm{s},\mathrm{r}_j}\right|^2\right\}}, \tag{4.1}$$

$$P_{\mathrm{out}} \leq P_{\mathrm{out}}^{\mathrm{thr}}, \tag{4.2}$$

where $u\left(\mathcal{R}, \mathbf{p}\right)$ represents a global objective function, E_{s} and E_{r_j} are the residual energy of the source and potential relay nodes, respectively, $P_{\mathrm{out}}^{\mathrm{thr}}$ is the threshold value of the outage probability, and $\mathbf{p} = \left[p_{\mathrm{s}}^{\mathrm{co},1}, p_{\mathrm{s}}^{\mathrm{co},2}, p_{\mathrm{r}_1}^{\mathrm{co},2}, \ldots p_{\mathrm{r}_{N-1}}^{\mathrm{co},2}\right]$ is the transmission power vector. The solution of (4)-(4.2) guarantees the participation of a proper number of nodes and the proper power allocation among these nodes. Actually, the energy-constrained nodes prefer to consume as little energy as possible, subject to the constraint on the desired outage performance, i.e., constraints (4.1) and (4.2). Hence, the transmission power vector is dependent on the potential relay nodes' decisions, i.e., $\mathbf{p}\left(\mathcal{R}\right)$. Once $\mathcal{R}$ is determined, $\mathbf{p}$ can be obtained by transforming (4.1) and (4.2) into equality constraints. That is, we can rewrite $u\left(\mathcal{R},\mathbf{p}\right)$ as $u\left(\mathcal{R}\right)$, following this convention below. However, due to the distributed and selfish features of the nodes, $u\left(\mathcal{R}\right)$ can not be easily evaluated by any one node, and hence an appropriate substitute must be found.

3   Correlated Equilibrium-Based Cooperation Policy Selection Game

The nodes in ad hoc networks are distributed and selfish, hence, their actions are strictly determined by self interest. For the purpose of distributed operation, we transform the behavior of $u\left(\mathcal{R}\right)$ into the individual objective functions u_{r_j} in the following sense. On the one hand, any potential relay node may be turned into the source node at the next time. Therefore, it would like to help the current source node, and expects a favor in return. The residual energy of source node can be viewed as the return on the kindness of relay nodes. On the other hand, we take into account the self-concern of each potential relay node, and employ the additional energy cost for cooperative transmission to reflect this. Then, u_{r_j} can lead us to a cooperation policy selection game which is modeled as

$$G = \left\langle \mathcal{R}_p, \left\{\mathcal{A}_{\mathrm{r}_j}\right\}_{\mathrm{r}_j \in \mathcal{R}_p}, \left\{u_{\mathrm{r}_j}\right\}_{\mathrm{r}_j \in \mathcal{R}_p} \right\rangle, \tag{5}$$

where the components of the game are given in the list:

1. The set of potential relay nodes $\mathcal{R}_p$ is the set of players.
2. $\mathcal{A}_{\mathrm{r}_j} = \{0, 1\}$ is the set of relay decision strategies for player r_j. Specifically, if r_j chooses to take part in the cooperative transmission, i.e., $\mathrm{r}_j \in \mathcal{R}$, $\mathcal{A}_{\mathrm{r}_j} = 1$; otherwise, $\mathcal{A}_{\mathrm{r}_j} = 0$.
3. $u_{\mathrm{r}_j} : \mathcal{A} \to \mathbb{R}$ is the individual utility that maps the joint strategy spaces $\mathcal{A} = \mathcal{A}_{\mathrm{r}_1} \times \cdots \times \mathcal{A}_{\mathrm{r}_{N-1}}$ to the set of real numbers. More precisely, the individual utility function should comply with the transformation rule mentioned above, that is

$$u_{\mathrm{r}_j}\left(\mathcal{A}_{\mathrm{r}_j}, \mathcal{A}_{-\mathrm{r}_j}\right) = E_{\mathrm{s}} - \frac{p_{\mathrm{s}}^{\mathrm{co},1} + p_{\mathrm{s}}^{\mathrm{co},2}}{2} - \mathcal{A}_{\mathrm{r}_j} \alpha_{\mathrm{r}_j} \frac{p_{\mathrm{r}_j}^{\mathrm{co},2}}{2E_{\mathrm{r}_j}}, \tag{6}$$

where $\mathcal{A}_{-\mathrm{r}_j}$ represents the joint strategies of the other players, and α_{r_j} is the pricing parameter of r_j, which weighs its cost compared to its reward. Also, $\mathcal{A} = \left(\mathcal{A}_{\mathrm{r}_j}, \mathcal{A}_{-\mathrm{r}_j}\right)$ is called a strategy profile.

Formally, the proposed game G can be expressed as

$$\max_{A_{\mathrm{r}_j} \in \mathcal{A}_{\mathrm{r}_j}} u_{\mathrm{r}_j} \left(A_{\mathrm{r}_j}, A_{-\mathrm{r}_j} \right), \quad \text{for all } \mathrm{r_j} \in \mathcal{R}_{\mathrm{p}}, \tag{7}$$

subject to constraints (4.1) and (4.2). Note that u_{r_j} is also a function of $A_{-\mathrm{r}_j}$, because it depends on the number of the relay nodes and the worst local CSI among the relay nodes, which are related to $A_{-\mathrm{r}_j}$.

In order to analyze the outcome of the proposed game G, we focus on an important generalization of the Nash equilibrium, known as the correlated equilibrium. For the distributed, competitive ad hoc network, the correlated equilibrium permits to coordinate the cooperation policy selection among potential relay nodes, hence, may lead to the most relevant noncooperative solution. There are several benefits for considering a correlated equilibrium, which are summarized in [6].

Definition 1. *Let $\Delta\mathcal{A}$ be the set of probability distributions on $\mathcal{A}$. A correlated strategy $P = (P(A))_{A \in \mathcal{A}} \in \Delta\mathcal{A}$ is a correlated equilibrium if for every strategy $A_{\mathrm{r}_j} \in \mathcal{A}_{\mathrm{r}_j}$ such that $P\left(A_{\mathrm{r}_j}\right) > 0$, and every alternative strategy $\tilde{A}_{\mathrm{r}_j} \in \mathcal{A}_{\mathrm{r}_j}$, it holds that*

$$\sum_{A_{-\mathrm{r}_j}} P\left(A_{\mathrm{r}_j}, A_{-\mathrm{r}_j}\right) u_{\mathrm{r}_j}\left(A_{\mathrm{r}_j}, A_{-\mathrm{r}_j}\right) \geq \sum_{A_{-\mathrm{r}_j}} P\left(A_{\mathrm{r}_j}, A_{-\mathrm{r}_j}\right) u_{\mathrm{r}_j}\left(\tilde{A}_{\mathrm{r}_j}, A_{-\mathrm{r}_j}\right). \tag{8}$$

P provides each player r_j with a private "recommendation" $A_{\mathrm{r}_j} \in \mathcal{A}_{\mathrm{r}_j}$, so as to allow a weak form of cooperation.

Theorem 1. *A correlated equilibrium always exists in the cooperation policy selection game G.* $\square$

Proof. The result from [9] shows that every finite game has a correlated equilibrium. Hence, Theorem 1 is justified, and enables the application of the proposed game. ∎

Theorem 2. *In the game G, if a CE achieves the highest social welfare, denoted as P^*, it is Pareto optimal.* $\square$

Proof. To find $P^*(\cdot)$, we first introduce an appropriate objective function, i.e.,

$$\max \sum_{A \in \mathcal{A}} P(A) \sum_{r_j \in \mathcal{R}_p} u_{\mathrm{r}_j}(A), \tag{9}$$

subject to constraint (8), and

$$P(A) \geq 0, \tag{9.1}$$

$$\sum_{A \in \mathcal{A}} P(A) = 1. \tag{9.2}$$

where (9) means that P^* is the solution to the highest social welfare, and constraints (8), (9.1) and (9.2) guarantee that P^* is the CE. If the resulting correlated equilibrium P^* is not Pareto efficient, there exists a different probability distribution $\tilde{P}$ such that $\sum_{A \in \mathcal{A}} \tilde{P}(A) u_{r_j}(A) \geq \sum_{A \in \mathcal{A}} P^*(A) u_{r_j}(A)$ for $\forall r_j \in \mathcal{R}$, and $\sum_{A \in \mathcal{A}} \tilde{P}(A) u_{r_j}(A) > \sum_{A \in \mathcal{A}} P^*(A) u_{r_j}(A)$ for some r_j, thus resulting in a higher value for the expected sum of utilities which contradicts the fact that P^* is the optimal solution to (9). This completes the proof. ∎

P^* can be obtained by linear programming method. However, all information is required to be available for optimization. The requirement is not possible for distributed r_j.

4   Distributed Algorithm for Cooperative Policy Selection

4.1   Algorithm Description

In this section, we present a distributed algorithm based on the regret matching procedure of [10] to obtain the set of correlated equilibria. Suppose that the proposed game G is played repeatedly through time: $n = 1, 2, \ldots$. At time $n+1$, given a history of play $h^n = (A^\tau)_{\tau=1}^n \in \prod_{\tau=1}^n \mathcal{A}$, each potential relay node $r_j \in \mathcal{R}_p$ chooses $A_{r_j}^{n+1} \in \mathcal{A}_{r_j}$ according to the average regret at time n. Then, the cooperative policy selection algorithm is executed independently by each potential relay node and summarized as follows.

1. **Initialization:** At the initial time $n = 1$, the source node calculates the minimum transmission power $p^{\mathrm{co},2}$ which satisfies constraint (4.2) and broadcasts the value to each potential relay node. Each potential relay node knows its CSI with the source node and initializes its strategy $A_{r_j}^1 \in \mathcal{A}_{r_j}$ arbitrarily.
2. **Iterative Update Process:** At the time n, each potential relay node r_j chooses a strategy $i \in \mathcal{A}_{r_j}$, and informs the source node of its choice. Then, the source node broadcasts the information which includes the number of relay nodes $n_t - 1$ and the worst-off CSI with relay nodes, i.e., $\omega^n = \min_{r_j \in \mathcal{R}} \left\{ d_j^{-\delta} |h_{s,r_j}|^2 \right\}$.
   - **Utility Update:** Each potential relay node r_j calculates its utility $u_{r_j}(A^n)$ according to (6), where $p_s^{\mathrm{co},2} = p_{r_j}^{\mathrm{co},2} = p^{\mathrm{co},2}/n_t$, and $p_s^{\mathrm{co},1}$ is the minimum value which satisfies constraint (4.1). Similarly, r_j calculates the utility for choosing the different strategy $k \in \mathcal{A}_{r_j}$.
   - **Average Regret Update:** If r_j replaces strategy i, every time that it was played in the past, with the different strategy $k \in \mathcal{A}_{r_j}$, the resulting difference in r_j's average utility up to time n is

$$D_{r_j}^n(i,k) = \frac{1}{n} \sum_{\tau \leq n : A_{r_j}^\tau = i} \left[u_{r_j}\left(k, A_{-r_j}^\tau\right) - u_{r_j}(A^\tau) \right]. \tag{10}$$

$D_{r_j}^n(i,k)$ represents the average regret at time n for not having played, every time that i was played in the past, the different strategy k.

- **Strategy Update:** According to the resulting average regret, r_j updates its relay decision strategy at the time $n + 1$:

$$A_{r_j}^{n+1} = \begin{cases} i, & D_{r_j}^n (i, k) \leq 0 \\ k, & others \end{cases} \quad (11)$$

Notes and Comments.

1. The proposed algorithm has low complexity. At each iteration, each r_j performs one table lookup to calculate its utility, two additions and two multiplication to update its regret value, and one comparison to determine the next strategy.
2. The proposed algorithm does not need r_j to know the individual strategies and utilities of other nodes, the global network structure, etc. This accords with the distributed characteristics of ad hoc networks.
3. We expand the applications of correlated equilibrium to the cooperation policy selection of ad hoc networks. Compared to [6, 7], we integrate the transmit power strategy selection into the algorithm, and modify the strategy update process in accordance with the specific space of two strategies, which avoids the bad convergence of fewer strategies.

4.2   Convergence Analysis

Let $z^n \in \Delta\mathcal{A}$ be the empirical distribution of play to time n, which can be viewed as an average or moving average frequency of play and given by

$$z^{n+1} = z^n + \frac{1}{n+1} \left(\mathbf{e}_{A^{n+1}} - z^n \right). \quad (12)$$

where $\mathbf{e}_{A^{n+1}} = [0, 0, \ldots, 1, 0, \ldots, 0]$ is the $|\mathcal{A}|$ dimensional unit vector with the one in the position of A^{n+1}.

Theorem 3. *If every potential relay node follows the proposed algorithm, the empirical distributions of play z^n converge almost surely as $n \to \infty$ to the set of correlated equilibria of the cooperation policy selection game G.* $\quad\square$

The proof that z^n converges to the set of correlated equilibria is presented in [6] and [10] respectively. Here, we only summarize and compare the two proofs.

1. In [6], the proof is based on a stochastic approximation convergence proof. A continuous time random process $z^n (t)$ is constructed by interpolating z^n. The tail behavior of the sequence $\{z^n\}$ is captured by the behavior of $z^n (t)$ for large t. Moreover, the trajectory of $z^n (t)$ converges almost surely to a trajectory whose dynamics are given by a different inclusion. Then, the asymptotically stable properties of the different inclusion tell us the tail behavior of $\{z^n\}$.
2. In [10], the proof relies on a recursive formula for the distance of the vector of regrets to the negative orthant. Particularly, in order to satisfy the conditions of Blackwell's approachability theorem, a multi-period recursion, where a large block of periods is combined together, substitutes for a one-period recursion.

5   Simulation Results and Analysis

In this section, we conduct simulations to study the performance of the proposed scheme over an energy-constrained cooperative ad hoc network. For both the local broadcasting and the long-haul cooperative channels, Rayleigh fading coefficients are modeled as unitary power, complex Gaussian random variables. The constant κ is set to 1, and the path loss exponent δ is set to 3. The Gaussian noise variance σ^2 is 10^{-12}W, and the outage capacity $\mathcal{C}_{out}$ is 1.4bps/Hz. The threshold value of outage probability is 10^{-4}. Besides, the pricing parameters of all nodes are set as one, but they have their own different residual energy.

Fig. 1 plots the evolution of regret value of worst player, when there exist 5, 10, 15 and 20 potential relay nodes, respectively. No matter how many potential relay nodes are placed in the ad hoc network, the correlated equilibrium can be obtained via using the proposed algorithm. From Fig. 1, we can find that: i) the individual regret value depends on not only its own strategy, but also the strategies chosen by other potential relay nodes, hence it can reflect the global convergence performance; ii) the more the potential relay nodes, the slower the convergence speed.

Fig. 2 presents the max-min fairness (see equation (4) subject to (4.1) and (4.2)) with different long-haul transmission distances. We exploit three algorithms which are plotted as reference, i.e., exhaustive search among all possible relay node combinations (Algorithm 1), complete cooperation (Algorithm 2) and complete noncooperation (Algorithm 3) among all potential relay nodes. Although Algorithm 1 can achieve the best cooperation policies from a max-min fairness perspective, its complexity is exponential. Our algorithm can obtain the same max-min fairness as Algorithm 1 with small complexity. In Algorithm 2, all potential relay nodes contribute their residual energy. Indeed, energy consumption caused by increasing the distance is averaged among all the nodes. However, it is disadvantageous for the node which has little residual energy with increasing the distance, hence, the max-min fairness becomes poor. In Algorithm 3, the

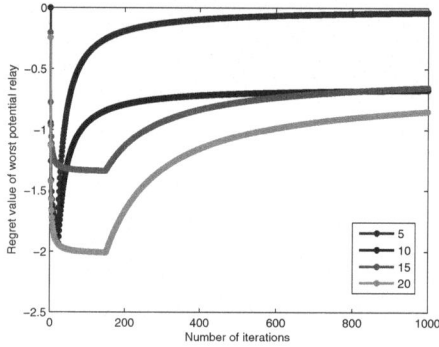

Fig. 1. Evolution of regret value of worst player for different number of potential relay nodes

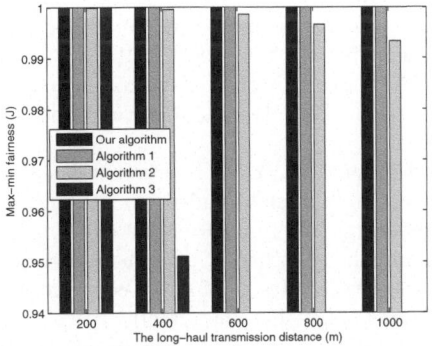

Fig. 2. Max-min fairness vs. the long-haul transmission distance in different algorithms

source node consumes much more energy than the other algorithms. Hence, it is more possible that the source node drains its energy firstly, while the other nodes have much retaining energy.

6   Conclusions

In this work, we design a cooperation policy selection scheme to achieve the global max-min fairness in terms of energy efficiency with outage performance constraint in energy-constrained cooperative ad hoc networks. Specifically, we model a cooperation policy selection game and focus on the CE to analyze the proposed game. Moreover, we develop an algorithm based on the regret matching procedure to obtain the correlated equilibrium. From the resulting correlated equilibrium, we can determine the proper cooperation policy. Both the theoretical analysis and simulation results demonstrate the efficiency of the proposed scheme.

Acknowledgement. This work is supported by the NSF of China (Grant No. 60972051, 61001107), the Major National Science & Technology Specific Projects (Grant No. 2010ZX03006-002-04), and the Open Research Fund of National Mobile Communications Research Laboratory, Southeast University (Grant No. 2010D09).

References

1. Wang, X., Vasilakos, A., Chen, M., Liu, Y., Kwon, T.: A survey of green mobile networks: Opportunities and challenges. ACM/Springer Mobile Networks and Applications (2011), doi:10.1007/s11036-011-0316-4
2. Dai, L., Chen, W., Cimini Jr., L.J., Letaief, K.B.: Fairness improves throughput in energy-constrained cooperative ad-hoc networks. IEEE Transactions on Wireless Communications 8(7), 3679–3691 (2009)
3. Qu, Q., Milstein, L.B., Vaman, D.R.: Cooperative and constrained MIMO communications in wireless ad hoc/sensor networks. IEEE Transactions on Wireless Communications 9(10), 3120–3129 (2010)

4. Sergi, S., Pancaldi, F., Vitetta, G.M.: A game theoretical approach to the management of transmission selection scheme in wireless ad hoc networks. IEEE Transactions on Communications 58(10), 2799–2804 (2010)
5. Sergi, S., Vitetta, G.M.: A game theoretical approach to distributed relay selection in randomized cooperation. IEEE Transactions on Wireless Communications 9(8), 2611–2621 (2010)
6. Krishnamurthy, V., Maskery, M., Yin, G.: Decentralized adaptive filtering algorithms for sensor activation in an unattended ground sensor network. IEEE Transactions on Signal Processing 56(12), 6086–6101 (2008)
7. Maskery, M., Krishnamurthy, V., Zhao, Q.: Decentralized dynamic spectrum access for cognitive radios: cooperative design of a non-cooperative game. IEEE Transactions on Communications 57(2), 459–469 (2009)
8. Han, Z., Pandana, C., Liu, K.: Distributive opportunistic spectrum access for cognitive radio using correlated equilibrium and no-regret learning. In: WCNC. IEEE (2007)
9. Hart, S., Schmeidler, D.: Existence of correlated wquilibria. Mathematics of Operations Research 14(1), 18–25 (1989)
10. Hart, S., Mas-Colell, A.: A simple adaptive procedure leading to correlated equilibrium. Econometrica 68(5), 1127–1150 (2000)

Energy- and Spectral-Efficient Wireless Cellular Networks

Mustafa Ismael Salman, Chee Kyun Ng, and Nor Kamariah Noordin

Department of Computer and Communication Systems Engineering,
Faculty of Engineering, University Putra Malaysia,
UPM Serdang, 43400 Selangor, Malaysia
mofisml3@yahoo.com, {mpnck,nknordin}@eng.upm.edu.my

Abstract. The limited spectrum resources and the negative impacts of carbon dioxide emission resulted from inefficient use of wireless technologies have led to the development of green radio. Both the energy and spectral efficiencies should be considered together to meet green radio requirements. In this paper, we investigate the trade-off between energy efficiency and spectral efficiency through different approaches. Cognitive radio is a paradigm-shift technology which is used to increase both the energy and spectral efficiencies. Some efficient spectrum sensing techniques are considered in terms of energy and time consuming. Furthermore, it can be shown that the power control strategies can play a key role in avoiding interference between cognitive and primary users, and hence it can also enhance both the energy and spectral efficiencies. In addition to cognitive radio, a new infrastructure for deploying the cellular base stations which is a heterogeneous infrastructure of macro-, pico-, and femto-cells is proposed to overcome the energy and bandwidth constraints. Further details related to hardware-constraints in a green base station have also been covered.

Keywords: Green radio, energy efficiency, spectral efficiency, cognitive radio, spectrum sensing, transmit power control, heterogeneous networks.

1   Introduction

The continuous rapid growth in wireless applications, devices and demands has led to a rapid growth in energy consumption and spectrum utilization. Due to this growth, both energy and bandwidth resources became so limited for wireless traffic. The limitation of energy resources is represented by the excessive emission of carbon dioxide (CO2) which is the chief greenhouse gas that results from wireless applications and other human activities and causes global warming and climate changes. This gas is accelerating continuously, as shown in Fig. 1, and need to be stabilized [1].

In the field of communications, more than 12,000 new base stations are installed every year to provide services to 300-400 million new subscribers around the world [2]. Many of these stations are driven by inefficient diesel generators which produces

Joel J.P.C. Rodrigues et al.: (Eds.): GreeNets 2011, LNICST 51, pp. 171–185, 2012.
© Institute for Computer Sciences, Social Informatics and Telecommunications Engineering 2012

the carbon footprint. Each base station antenna consumes an average power of 1KW which means 8,800 KWh each year [3]. A network with a medium size normally consists of 12-15,000 cell sites, each can serve two technologies (2G & 3G), and each technology needs around three antennas per technology, which tends to a total energy of 736,000 MWh which can run 168,000 European family houses [3]. Such statistics give a clear indication that the Information and communication technologies (ICT) contribute in the total world's carbon footprint.

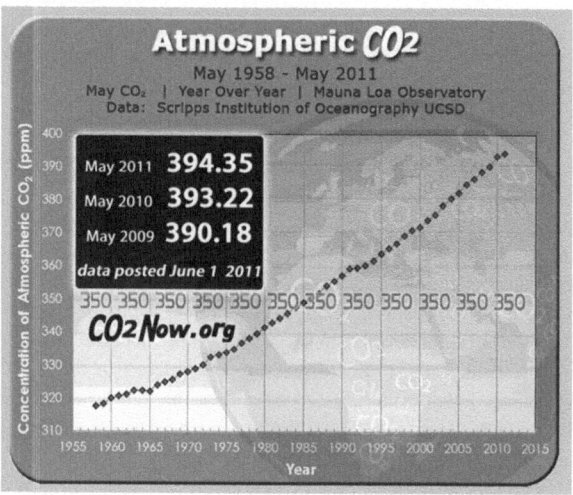

Fig. 1. Concentration of atmospheric CO2

On the other hand, the limitation in bandwidth resources is represented by the fact that the spectrum is not free and it is fixed. Data traffic has increased in the recent years due to the presence of iPhone and other smart software technologies and due to the variety of applications, and it is expected to grow more with the introduction of LTE-A which supports 100Mbps for down-link. In order to achieve such high bit rate, we need to improve the spectral efficiency of the channels.

So far, the improvement in spectral efficiency has been the main interest of the research without much consideration of the energy efficiency metrics. Those two parameters (energy- and spectral-efficiency) should be considered together in order to meet what is known as "Green Radio". In Green radio, both energy efficiency and spectral efficiency need to be maximized. However, they are, sometimes, two conflicting parameters which mean that any increase in spectral efficiency will lead to undesirable increase in power consumption [4]. Therefore, finding a trade-off between the energy- and spectral-efficiency is the goal of this paper. In this paper, we will survey and propose the state-of-the-art the energy- and spectral- efficient technologies that need to be considered in the current and next generation wireless networks.

The rest of the paper is organized as follows. In Section 2, both the energy and spectrum efficiencies are defined and the trade-off between those two parameters is investigated. Section 3 shows the green wireless base stations architecture. Cognitive radio as a promising technology for green radio will be covered in section 4. In Section 5, Heterogeneous network optimization will be suggested to meet a green next generation wireless communication deployment. Finally, the conclusions and recommendations for future work are drawn in section 6.

2   The Trade-Off between Energy and Spectral Efficiencies

The more energy-efficient communication system, the less energy required to achieve the same task. On the other hand, the more bandwidth- (spectral-) efficient communication system, the more bits per second it can transfer through the same channel. The maximization of spectral efficiency is one of the main targets that should be achieved in the next generation networks. Fig. 2 shows a comparison between spectral efficiencies required in different technologies for down-link and up-link transmission.

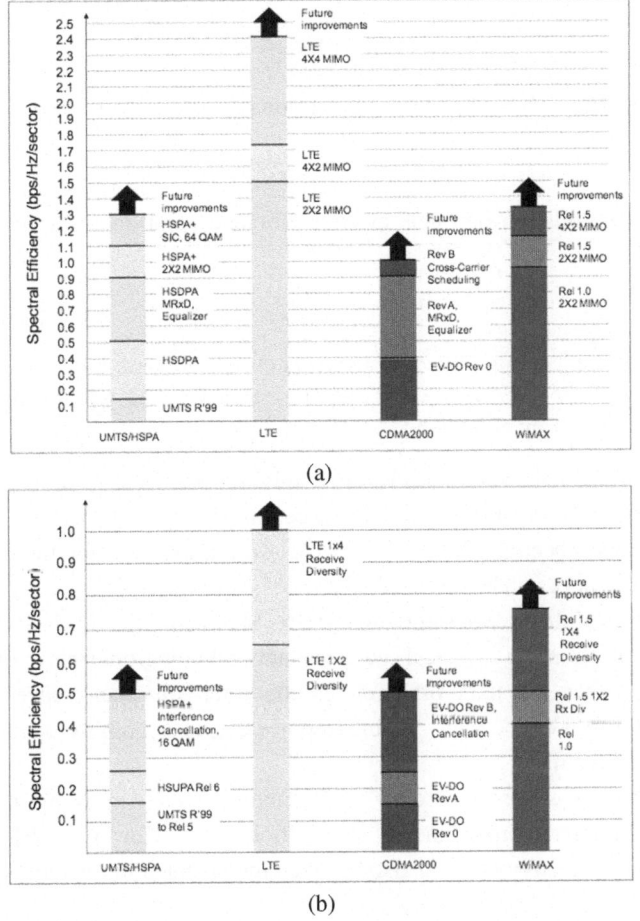

(a)

(b)

Fig. 2. Spectral efficiency comparison; (a) downlink and (b) uplink

On the contrary, there was no big interest in improving the energy efficiency. Now, green communication is a major challenge and introducing energy-efficient communication systems should be added to the list of major interests of academic and industrial researches. Sometimes, energy and spectral efficiency conflict each other. To formulate this trade-off, the Shannon's capacity equation for point to point communication with AWGN plays the key role [5]:

$$C = B \, log_2 \left(1 + \frac{S}{N}\right) \quad bps \, . \tag{1}$$

where B is the channel bandwidth (Hz), S is the signal power, and N is the noise power which is NoB, where N_o is the power spectral density for AWGN. Therefore:

$$C = B \, log_2 \left(1 + \frac{S}{N_o B}\right).$$

The bandwidth (spectral) efficiency (η_B), is the achievable transmission rate per unit bandwidth (bps/Hz):

$$\eta_B = log_2 \left(1 + \frac{S}{BN_o}\right). \tag{2}$$

The energy efficiency (η_E), is the transmission rate per unit energy (bps/W):

$$\eta_E = B log_2 \left(1 + \frac{S}{BN_o}\right)/S \, . \tag{3}$$

Therefore, the relation between spectral efficiency and energy efficiency is shown below [7]:

$$\eta_E = \frac{\eta_B}{(2^{\eta_B} - 1)N_o} \, . \tag{4}$$

This relation can be represented by a convex curve shown in Fig. 3.

However, there are many other factors that affect the relation between spectral efficiency and energy efficiency:

- Physical layer transmission: The transmission parameters and strategies, such as modulation order, transmission distance and coding scheme, may also affect the energy and spectral efficiencies. An extensive analysis for the energy-efficient transmission in wireless networks has been achieved in [5].
- Multi-cell/Multi-user systems: the above Shannon equation is for point-to-point communication. In the real wireless network environments many parameters should affect the trade-off such as the inter-cell interference and inter-user interference. i.e., the inter-cell interference degrades both the spectral and energy efficiencies [6].
- The channel state: According to Shannon's equation, using high bandwidth channels can improve the energy efficiency. However, delay spread and frequency selectivity should also be taken into account in trade-off calculation [5].

Hardware energy consumption: circuit power and real practical hardware constraints should be considered in the calculation of this trade-off:

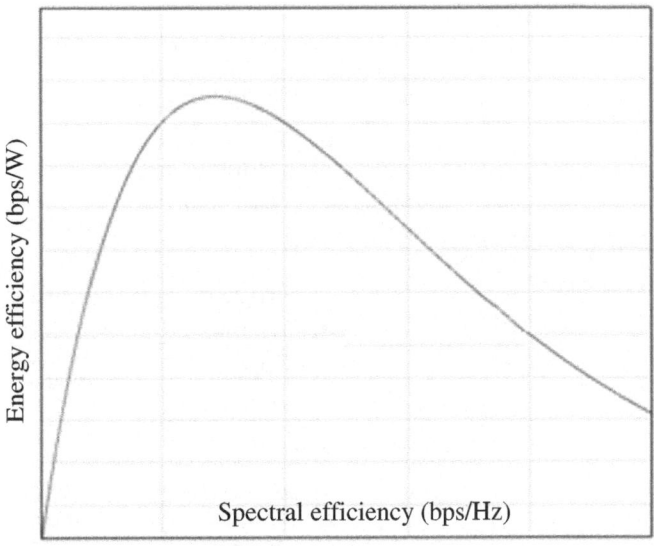

Fig. 3. Energy vs. spectral efficiency in point-to-point communications

3   Green Base Station

Due to the increase in wireless demands and users, the cellular operators had to increase the number of base stations to provide more wireless services to bigger number of users. This growth in number of base stations makes the operators looking seriously for efficient equipments to overcome the energy constraints. The cellular base station consists of several power consuming equipments as shown in Fig. 4. Those equipments consume different amount of power in different technologies as shown in Table 1:

Table 1 shows that the power amplifier is the major source of power consumption in the base station. The more energy-efficient base station, the less heat produced by the equipment, and thus, the less amount of air-conditioning required for cooling. Therefore, the improvement of efficiency of the power amplifier will reduce the power consumption of the main parts in a base station. The energy efficiency of the power amplifier can be improved by using a proper linearization and DSP methods to decrease the required linear area. Besides the power amplifier improvement, there are several approaches that have been proposed in the literature to introduce more energy-efficient base stations. [7, 8] proposed a dynamic planning based on traffic intensity by switching off the underutilized base stations (i.e. during night periods) while maintaining the required quality of service. The results show that the implementation of this approach can save up to 50% of the power consumption. The cooperation between two networks by switching off one of them during low traffic has been investigated in [9] in terms of power saving.

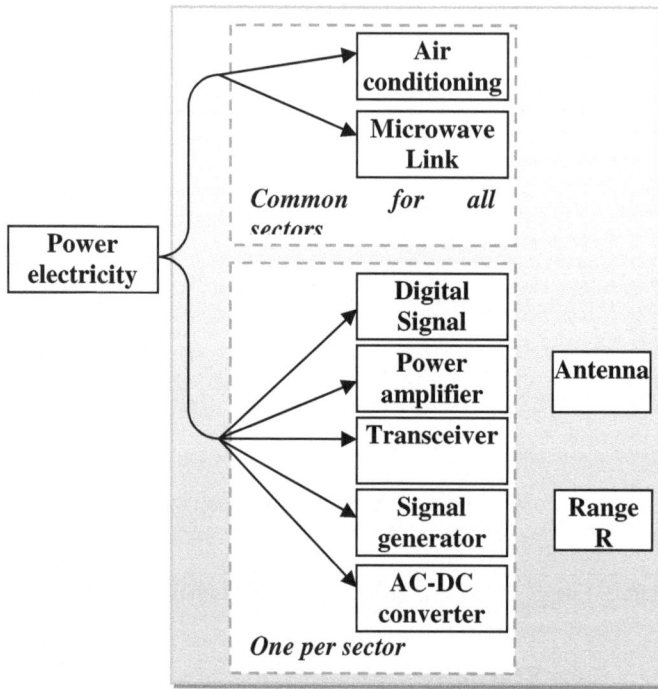

Fig. 4. Power consuming equipments in cellular base station

Table 1. Power consumption of different parts of wireless base stations

Equipment	WiMAX	HSPA	LTE
Digital signal processing	100 W	100 W	100 W
Power amplifier SISO (1x1)	100 W 10% 40dbm	300 W 6.67% 43dbm	350 W 6.3% 43 dbm
Power amplifier MIMO	10.4 W 11.54% 30 dbm	10.4 W 11.54% 30 dbm	10.4 W 11.54% 30 dbm
Transceiver	100 W	100 W	100 W
Signal generator	384 W	384 W	384 W
AC-DC converter	100 W	100 W	100 W
Air conditioning	690 W	690 W	690 W
Microwave link	80 W	80 W	80 W

Also, multiple-input and multiple-output, or MIMO, is considered in the new transmitting systems to improve the system capacity. By improving the spectral efficiency, the transmission duration is reduced which tends to reduction in transmitted power and circuit power consumption. On the other hand, more active components are needed by exploiting MIMO which increase the total power consumption. According to these conflicting facts, the impacts of MIMO techniques on energy efficiency has been addressed in [10, 11] and it was shown that cooperative MIMO transmission and reception can outperform the SISO systems in terms of the energy efficiency, as shown in Fig. 5, when the adaptive modulation is used to control the transmit and circuit energy consumption [10]. However, SISO systems is more energy-efficient than MIMO systems when the latter is not combined with adaptive modulation. Further investigation is required to optimize the MIMO systems for next generation wireless networks in terms of energy efficiency, spectral efficiency and overall complexity.

4   Cognitive Radio

The spectral utilization is one of the most critical problems that face the rapid developments of wireless communications. Previous researches followed some approaches to increase the spectral efficiency at the expense of energy efficiency. Recent researches in cognitive radio technology [4, 12-14] have brought a significant improvement towards green wireless communication. The cognitive radio technique will provide the wireless users with a high bandwidth and allow them to use the unutilized (white) spectrum through dynamic spectrum access techniques [15].

In order to achieve efficient utilization of the spectrum, the unlicensed cognitive radio user (secondary user) can adapt its transmission and reception parameters to avoid interference with the primary user, and thus, it gives a significant enhancement

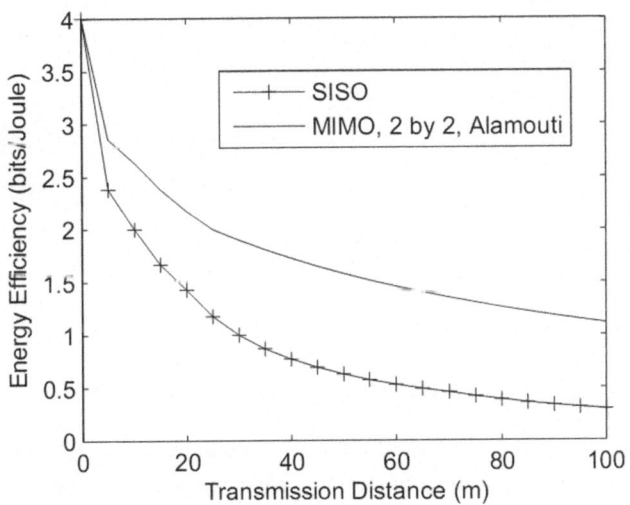

Fig. 5. Comparing energy efficiency of MIMO and SISO

to green wireless networks. In [12], an energy optimization framework shown in Fig. 6 has been proposed to adjust parameters (e.g., modulation, radiated power and coding) and components characteristics (e.g. power amplifier) using cognitive radio.

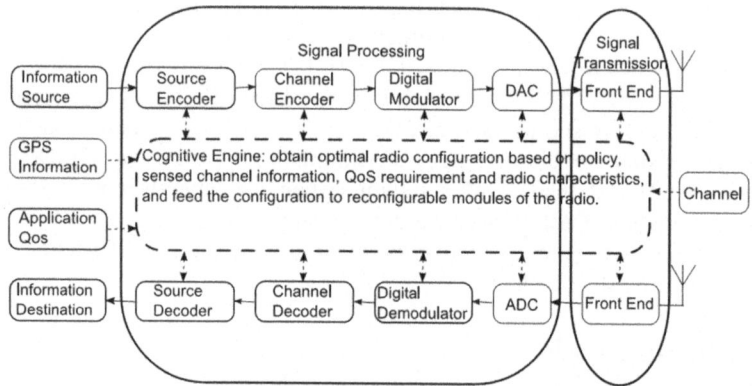

Fig. 6. Cognitive radio energy optimization framework

According to [12], Fig. 7 shows the simulation of this framework using cognitive transmission over adaptive modulation along with power amplifier radiated power and it shows a significant energy saving up to 75%. However, the energy saving becomes less significant as the distance increases.

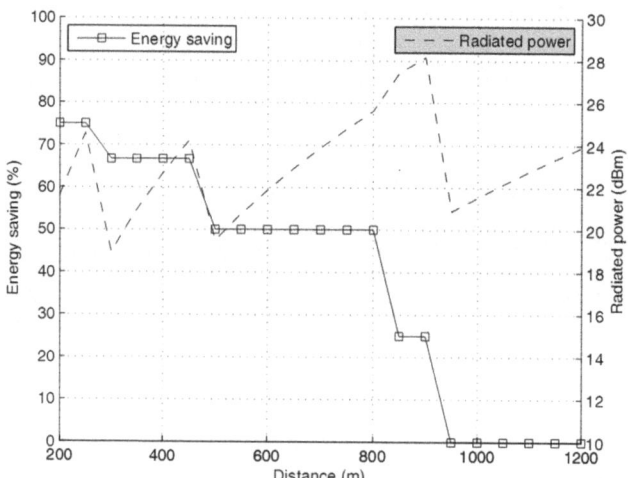

Fig. 7. Energy saving with cognitive transmission

The cognitive radio is a paradigm-shift technology that let the user interact intelligently with the environment through what is known as cognition cycle. This cycle consists mainly of four consecutive steps as shown in Fig. 8.

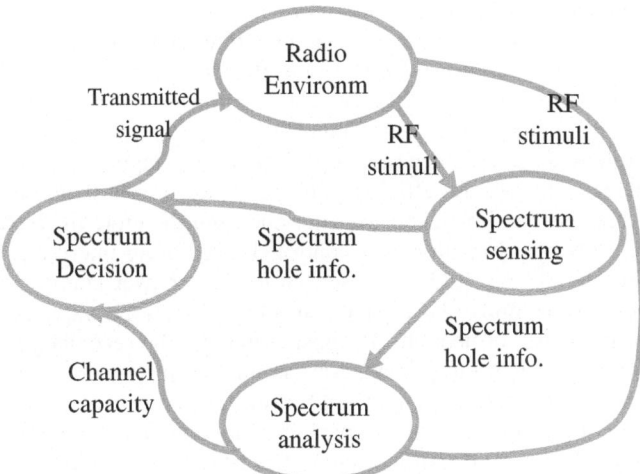

Fig. 8. Cognition cycle

First, the cognitive radio technique will let the unlicensed users to determine which portion of the spectrum is currently not used (spectrum holes) and detect the existence of the primary licensed users, (spectrum sensing) [16]. Those unlicensed users should keep monitoring the spectrum continuously and therefore, they will still active. Due to this pivot role, the spectrum sensing is considered as one of the most time and energy consuming part of the cognitive radio device. Previous work concentrated on the time overhead of the spectrum sensing [17, 18]. In [16], an optimal sensing duration has been designed to maximize the throughput using the energy detection scheme. [17] studied the trade-off between the spectrum usage time and the energy efficiency of the spectrum sensing. Recently, and due to the green communications trends, the energy consumption of the spectrum sensing becomes one of the most challenges that face the academic researches nowadays. J. Wei and X. Zhang proposed an energy-efficient spectrum sensing technique using cluster-and-forward based Distributed Spectrum Sensing (DSS) [19]. This technique has shown a significant decrement in the total energy consumption while maintaining high sensing accuracy. A further improvement to the energy efficiency of the spectrum sensing has been proposed by [20]. The researchers proposed a Time-Division Energy-Efficient (TDEE) sensing technique that well balanced the trade-off between spectral efficiency and energy consumption by investigating heterogeneous and homogeneous networks. Although there was a good investigation for the efficient spectrum sensing, the green cognitive radio still need more interest to study the trade-off between all the spectrum sensing parameters which are: energy efficiency, spectral efficiency, sensing time and accuracy. However, the complexity of spectrum sensing can be reduced by exploiting some artificial intelligent techniques [14].

After sensing the spectrum, the cognitive radio user has to select the best available channel to meet the quality of service requirement over all available spectrum bands (spectrum management) [21]. Then, it allows the secondary user to access this channel along with the other users (spectrum sharing).

For cognitive radio, the introduced solutions for spectrum sharing can be classified into three phases: i.e., according to their architecture, spectrum allocation behavior, and spectrum access technique [22]. Upon the information available from the spectrum sensor, the cognitive radio user varies its transmitted power to maximize its performance. This operation is called "Transmit Power Control (TPC)". TPC can play an important role in terms of green radio optimization by improving power efficiency. In order to increase the spectral efficiency, higher power levels should be allocated to more fading channels and low levels to better ones and therefore the interference will be minimized. Previous work in power control showed a big interest on maximizing the spectral efficiency, e.g., [23] proposed an optimal power control over different fading channels to maximize the ergodic capacity of the secondary user taking into account the primary user protection. By considering that the secondary user can share the licensed spectrum with the primary user as long as its interference power to the primary user still below a specific threshold level, [24] investigates the capacity gain offered by this spectrum sharing approach in Rayleigh fading environments, and derived an optimal power allocation scheme from the outage and ergodic capacities points of view. However, we can improve the power efficiency by using power truncation such that the secondary user can transmit in good channel conditions and abstain from transmission otherwise [25]. In such cases, bad channel conditions will cause long time delays which improve the power efficiency due to power truncation but, on the other hand, will minimize the spectral efficiency. Therefore, a trade-off between a spectral efficiency and power efficiency has been investigated in [25] as shown in Fig. 9. This figure shows the energy efficiency and goodput versus peak power in case of power truncation. Here, we can see that, at low level of peak power, the goodput is increasing as the peak power decreases. Further decreasing in peak power will result in goodput reduction while the energy efficiency is still increasing.

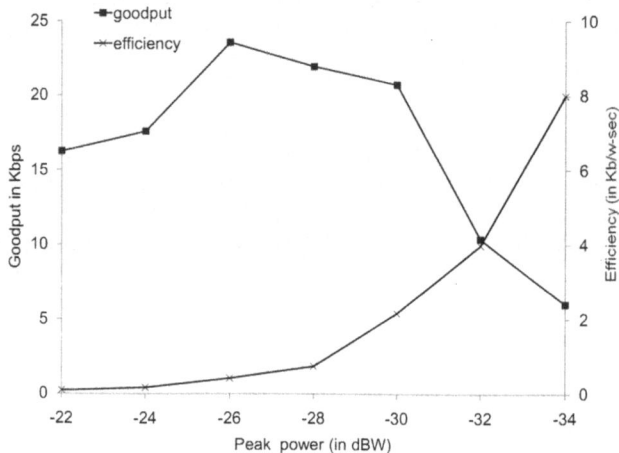

Fig. 9. Goodput & efficiency vs. P_{peak} at $I_{max} = 2 \times 10^{-12}$W and SNR=12dB

When the primary user has been detected, the cognitive radio user should leave the channel and access other unutilized channel (Spectrum mobility). Spectrum mobility presents a new type of handoff in next generation networks which is known as spectrum handoff. The purpose of spectrum mobility is to ensure a smooth and quick transition between different operational modes [15].

5   Heterogeneous Networks

Current wireless cellular networks are homogeneous networks that are deployed using a macro-centric planned process. In such deployment, all macro base stations have similar parameters, e.g., backhaul connectivity to the data network, antenna patterns, power levels, and receiver noise floor [26]. With the growing of wireless traffic demands, more flexible deployment architecture is needed to overcome capacity, and link budget limitations and to maintain user satisfaction. Therefore, the heterogeneous network is the alternative deployment that brings the network closer to the user. The heterogeneous network consists of a high power macro base station, several low power pico, femto, and/or relay base stations. Although the major target of heterogeneous network deployment is to improve capacity, it has a significant potential to improve the energy efficiency. [27] studied the power consumption of combined macro- and femto-cells architecture and found that the overall power consumption can be reduced depending on the uptake of femto-cell usage. Different femtocell capacities have been used in the simulation made by [27] as shown in Fig. 10.

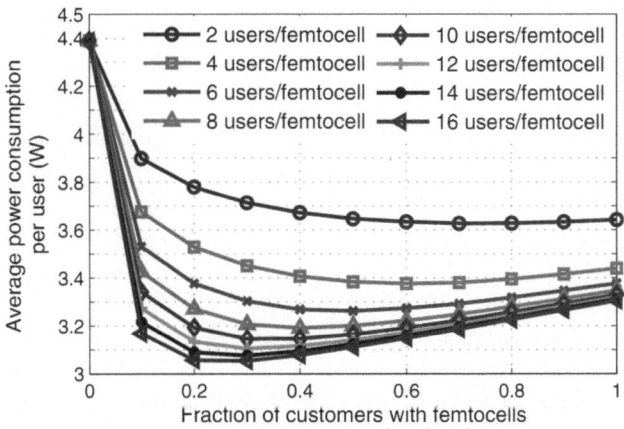

Fig. 10. Power consumption per user for different levels of femtocell support

It is clear from the figure that more power can be saved as the capacity of femtocell increases. This result can be explained from two points of view. First, the smaller the capacity of femtocell the more users need to connect with macrocell, and therefore, the more power required to perform this long connection. Second, the more users served by one active femtocell, the more other femtocells can be switched to sleep mode, and therefore, the more power can be saved. Also, a combination of pico- and macro-cells architecture has been investigated by [28] and a simulation has been made to show that such deployment can reduce the energy consumption for high data rate user demands as shown in Fig. 11.

In the mentioned work, the researchers build their results upon the power consumption measurements. Other metrics like Energy Consumption Gain (ECG) and Energy Consumption Ratio (ECR) have been identified to quantify energy consumption performance in small cell size deployment [29]. Further verifications for power saving can be achieved by measuring the energy efficiency of such networks. In [30], the overall average cell energy efficiency with unit of bits per Joule has been derived for heterogeneous cellular deployment to as follows:

$$E_{e,cell} = \frac{R_{Ma} + \sum_{n=1}^{N} R_{pi,n}}{P_{Ma} + \sum_{n=1}^{N} P_{pi,n}}. \tag{5}$$

Where R_{ma} and $R_{pi,n}$ denotes the average data rate provided by macro and nth pico cells respectively. P_{Ma}, and $P_{pi,n}$ represents the power consumed by macro and nth pico stations respectively.

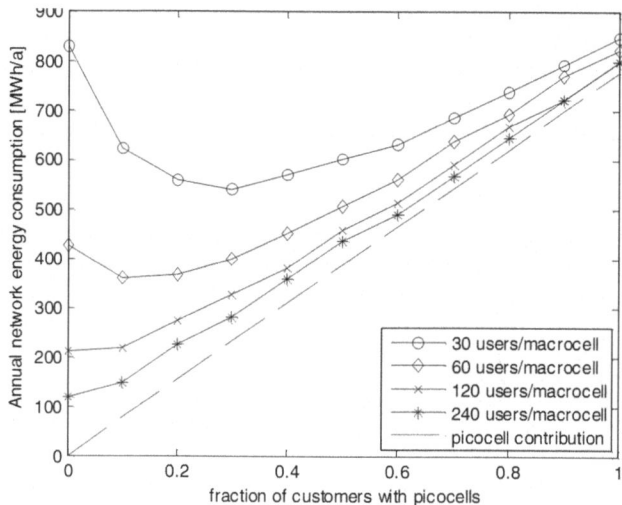

Fig. 11. Power consumption per user for different levels of femtocell support

In addition to the energy efficiency, [30] studied the effect of cell size on the energy efficiency performance by introducing another important metric which is: Area Energy Efficiency (AEE). Further improvement in terms of energy efficiency can be added to the heterogeneous cellular networks by considering low power sleep mode solution when a cell is not used [31]. Further green techniques, projects, and metrics have been surveyed and summarized in [32].

6   Conclusions

In this paper, the relation between energy efficiency and spectral efficiency is investigated under several approaches and constraints. Several approaches, such as cognitive radio and heterogeneous network, are discussed in term of energy and spectral efficiency optimization to meet green radio requirements. In cognitive radio, we show that it can improve the energy efficiency in spite of that its major role is to maximize spectral efficiency. Designing an efficient spectrum sensing algorithm in terms of energy, time and accuracy is one of our interests for future work. Also, heterogeneous wireless network can be used to maximize both the spectral and energy efficiency. Sleep-mode strategy and cell-size reduction are the main techniques used with deployment of heterogeneous network. In addition to that, green base station architecture is proposed to enhance the energy efficiency of energy consuming parts of the base station (e.g. power amplifier). And multiple antennas (MIMO) can be used to maximize the capacity and further investigation needed to optimize the design of MIMO transmitters and receivers in terms of energy efficiency, spectral efficiency and overall complexity.

References

1. CO2 Now | CO2 Home, http://co2now.org/
2. Sistek, H.: Green-tech base stations cut diesel usage by 80 percent. Green Tech - CNET News, http://news.cnet.com
3. Amanna, A.: Green Communications. Annotated Literature Review and Research Vision (2010)
4. Vo, Q.D., Choi, J.-P., Chang, H.M., Lee, W.C.: Green perspective cognitive radio-based M2M communications for smart meters. In: IEEE International Conference on Information and Communication Technology Convergence (ICTC), pp. 382–383. IEEE Press, Jeju (2010)
5. Miao, G., Himayat, N., Li, Y., Swami, A.: Cross-layer optimization for energy-efficient wireless communications: a survey. Wireless Communications and Mobile Computing 9(4), 529–542 (2009)
6. Guowang, M., Himayat, N., Li, G.Y., Koc, A.T., Talwar, S.: Interference-Aware Energy-Efficient Power Optimization. In: IEEE International Conference on Communications, ICC 2009, pp. 1–5. IEEE Press, Dresden (2009)
7. Marsan, M.A., Chiaraviglio, L., Ciullo, D., Meo, M.: Optimal Energy Savings in Cellular Access Networks. In: IEEE International Conference of the Communications Workshops, ICC Workshops, pp. 1–5. IEEE Press, Dresden (2009)
8. Chiaraviglio, L., Ciullo, D., Meo, M., Marsan, M.A.: Energy-efficient management of UMTS access networks. In: 21st IEEE International Conference on Teletraffic Congress, ITC 21, pp. 1–8. IEEE Press, Paris (2009)
9. Marsan, M.A., Meo, M.: Energy Efficient Management of Two Cellular Access Networks. SIGMETRICS Perform. Eval. Rev. 37(4), 69–73 (2010)
10. Shuguang, C., Goldsmith, A.J., Bahai, A.: Energy-efficiency of MIMO and Cooperative MIMO Techniques in Sensor Networks. IEEE Journal on Selected Areas in Communications 22(6), 1089–1098 (2004)
11. Wenyu, L., Xiaohua, L., Mo, C.: Energy efficiency of MIMO transmissions in wireless sensor networks with diversity and multiplexing gains. In: IEEE International Conference on Acoustics, Speech, and Signal Processing (ICASSP 2005), pp. 897–900. IEEE Press (2005)

12. An, H., Srikanteswara, S., Reed, J.H., Xuetao, C., Tranter, W.H., Kyung Kyoon, B., Sajadieh, M.: Minimizing Energy Consumption Using Cognitive Radio. In: IEEE International Conference on Performance, Computing and Communications Conference, IPCCC, pp. 372–377. IEEE Press, Austin

13. Palicot, J.: Cognitive radio: an enabling technology for the green radio communications concept. In: International Conference on Wireless Communications and Mobile Computing: Connecting the World Wirelessly. ACM, Leipzig (2009)

14. Grace, D., Jingxin, C., Tao, J., Mitchell, P.D.: Using Cognitive Radio to Deliver Green Communications. In: IEEE 4th International Conference on Cognitive Radio Oriented Wireless Networks and Communications, pp. 1–6. IEEE Press, Hannover (2009)

15. Akyildiz, I.F., Lee, W.-Y., Vuran, M.C., Mohanty, S.: NeXt Generation/dynamic Spectrum Access/cognitive Radio Wireless Networks: A Survey. Computer Networks 50(13), 2127–2159 (2006)

16. Shellhammer, S.J.: Spectrum Sensing in IEEE 802.22. IAPR Wksp. Cognitive Info. Processing (2008)

17. Ying-Chang, L., Yonghong, Z., Peh, E.C.Y., Anh Tuan, H.: Sensing-Throughput Tradeoff for Cognitive Radio Networks. IEEE Transactions on Wireless Communications 7(4), 1326–1337 (2008)

18. Su, H., Zhang, X.: Power-Efficient Periodic Spectrum Sensing for Cognitive MAC in Dynamic Spectrum Access Networks. In: IEEE Conference on Wireless Communications and Networking (WCNC), pp. 1–6. IEEE Press, Sydney (2010)

19. Jin, W., Xi, Z.: Energy-Efficient Distributed Spectrum Sensing for Wireless Cognitive Radio Networks. In: INFOCOM IEEE Conference on Computer Communications Workshops, pp. 1–6. IEEE Press (2010)

20. Liu, Y., Xie, S., Zhang, Y., Yu, R., Leung, V.: Energy-Efficient Spectrum Discovery for Cognitive Radio Green Networks. Mobile Networks and Applications, 1–11 (2011)

21. Budiarjo, I., Lakshmanan, M., Nikookar, H.: Cognitive Radio Dynamic Access Techniques. Wireless Personal Communications 45(3), 293–324 (2008)

22. Weiss, T.A., Jondral, F.K.: Spectrum pooling: an innovative strategy for the enhancement of spectrum efficiency. IEEE Communications Magazine 42(3), 8–14 (2004)

23. Rui, Z.: Optimal Power Control over Fading Cognitive Radio Channel by Exploiting Primary User CSI. In: IEEE Global Telecommunications Conference, IEEE GLOBECOM, pp. 1–5. IEEE Press, New Orleans (2008)

24. Musavian, L., Aissa, S.: Ergodic and Outage Capacities of Spectrum-Sharing Systems in Fading Channels. In: IEEE Global Telecommunications Conference, GLOBECOM 2007, pp. 3327–3331. IEEE Press (2007)

25. Tripathi, P.S.M., Cianca, E., di Sanctis, M., Ruggieri, M., Prasad, R.: Truncated Power Control Over Cognitive Redo Networks: Trade-off Capacity/Energy Efficiency. In: 13th International Symposium on Wireless Personal Multimedia Communications (WPMC), Recife, Brazil (2010)

26. Khandekar, A., Bhushan, N., Ji, T., Vanghi, V.: LTE-Advanced: Heterogeneous networks. In: IEEE European Wireless Conference (EW), pp. 978–982. IEEE Press (2007, 2010)

27. Ying, H., Laurenson, D.I.: Energy Efficiency of High QoS Heterogeneous Wireless Communication Network. In: IEEE Conference on Vehicular Technology Conference Fall (VTC 2010-Fall), pp. 1–5. IEEE Press, Ottawa (2010)

28. Claussen, H., Ho, L.T.W., Pivit, F.: Effects of Joint Macrocell and Residential Picocell Deployment on the Network Energy Efficiency. In: IEEE 19th International Symposium on Personal, Indoor and Mobile Radio Communications, pp. 1–6. IEEE Press, Cannes (2008)

29. Badic, B., O'Farrrell, T., Loskot, P., He, J.: Energy Efficient Radio Access Architectures for Green Radio: Large versus Small Cell Size Deployment. In: IEEE Conference on Vehicular Technology Conference Fall, pp. 1–5. IEEE Press, Anchorage (2009)
30. Wei, W., Gang, S.: Energy Efficiency of Heterogeneous Cellular Network. In: IEEE Conference on Vehicular Technology Conference Fall, pp. 1–5. IEEE, Ottawa (2010)
31. Haratcherev, I., Fiorito, M., Balageas, C.: Low-Power Sleep Mode and Out-Of-Band Wake-Up for Indoor Access Points. In: IEEE GLOBECOM Workshops. IEEE (2009)
32. Wang, X., Vasilakos, A.V., Chen, M., Liu, Y., Kwon, T.T.: A Survey of Green Mobile Networks: Opportunities and Challenges. Mobile Networks and Applications, 1–17 (2011)

Energy Packet Networks:
ICT Based Energy Allocation and Storage
(Invited Paper)

Erol Gelenbe

Intelligent Systems & Networks
Dept. of Electrical & Electronic Engineering
Imperial College, London SW7 2BT, UK
e.gelenbe@imperial.ac.uk

Abstract. In the presence of limitations in the availability of energy for data centres, especially in dense urban areas, a novel system that we call an *Energy Packet Network* is discussed as a means to provide energy on demand to Cloud Computing servers. This approach can be useful in the presence of renewable energy sources, and if scarce sources of energy must be shared by multiple computational units whose peak to average power consumption ratio is high. Such a system will use energy storage units to best match and smooth the intermittent supply and the intermittent demand. The analysis of such systems based on queueing networks is suggested and applied to a special case for illustration.

1 Introduction

The Cloud offers dynamic provisioning of computing and networking resources to applications that can thus be executed, on demand and in a distributed manner at the best possible cost and quality of service. Middleware can help users dynamically locate and select the most effective Cloud services that meet their needs, and Cloud service providers can compete among themselves to meet the users' needs in the most cost-effective way. While the concentration of Cloud services in data centres that can provide these services cheaply is an attractive option, such concentrated centres have overall energy needs that are often prohibitive [20]. In fact, when a data centre is installed its present and future energy needs have to be planned in advance.

Such installations are already becoming quite difficult to install in large urban areas in Europe such as Paris or London where the computational needs are greatest. Similarly, CO2 taxes are already deterring investments in such larger centres [41]. An alternative approach for large urban areas is to distribute computing power over a large number of smaller sites which will be operating asynchronously, and to dispatch *energy on demand* when it is required by computations, rather than to guarantee a high level of power availability all the time.

Joel J.P.C. Rodrigues et al.: (Eds.): GreeNets 2011, LNICST 51, pp. 186–195, 2012.
© Institute for Computer Sciences, Social Informatics and Telecommunications Engineering 2012

This leads us to the concept of *Energy Packet Networks* (EPN) which are integrated adaptive electrical energy storage, distribution and consumption systems that we proposed recently [35,43]. In addition to the conventional scheme for distributing energy based on instantaneous flow of current towards points of energy consumption, EPNs offer the smart request and dispatching of units of energy called *Energy Packets* (EP) to meet the demands of computing consumers such as Cloud users, as well as other electrical devices and appliances. In EPN, smart dynamic generation and storage combine with smart on-demand request and dispatching of electric power. Such systems are particularly well adapted to environments where renewable energy sources are common, and where effective means for storing energy, such as electric cars and uninterruptible power supplies, are available.

In the case of energy provisioning for the Cloud, such a system will include geographically distributed renewable energy sources, conventional backup sources of energy that originate from fossil and nuclear power plants, and a distribution network, together with distributed energy storage facilities and the small or large data centres that create demand and that are provisioned with energy on demand. Many of the consumption sinks may also be coupled at close distance with one or more storage facilities.

The flow of energy in the EPN will be controlled by *Smart Energy Dispatching Centres (SEDCs)* which receive requests from both the consumers (e.g. Cloud servers) and from storage centres that wish to be replenished. The SEDCS then optimise the energy flows by making the best use of renewable energy availability, storage centre state, and existing pricing policies, while satisfying the demands and minimising peak energy flows through buffering and scheduling. SEDCs will obviously be computer control centres which receive information and make dispatching decisions from/to to energy switches via data communication networks. The basic unit of energy in an EPN, an EP, can be viewed as a pulse of power that lasts a certain time; it constitutes the basic energy delivery unit of our system, say in KWH. The energy content of such a packet must be small enough to be close to the smallest energy needs of consumers, e.g. running a particular job or group of jobs at a data centre, and yet large enough to be measurable and billable as a significant and useful quantity.

Although we not detail the approach in this paper, our vision is that the SEDCs will use distributed adaptive schemes such as those developed for smart packet network routing [10,19]. Thus when a request for cloud computing arises, the SEDCs will not only select the computational resource which can do this at lowest energy and/or economic cost, but will also dynamically provision the energy dynamically. Similar ideas have been suggested for energy-aware network routing in recent papers [28,29].

2 A Model for EPNs

Although the system we consider is intuitively appealing, its value resides in its potential as a better means of delivering electric power for applications such as data centres and clouds whose energy consumption goes hand in hand with their computational activities. Thus we need to model EPNs and evaluate their performance advantages through system analysis. The paradigm of queueing networks is particularly favorable for studying such systems hus in this section we will develop a model representation for EPNs. The theoretical framework that we will use is a class of stochastic networks [2] known as G-networks [6,9], which:

- Represent the flow of a discretised commodity (the EPs) and its storage in storage centres (STs) represented by queues,
- Representing the choices that are made regarding the distribution of these flows through the EPN via routing probabilities for the EPs; note that these routing probabilities would normally be determined by the SEDCs and can be modified so as to optimise the system's performance,
- The model can incorporate the flow of data as described in [43] so that control decisions can be taken, for instance to select the demands made by storage units and Cloud servers.

The data flows themselves will be coupled to the decisions to transfer EPs towards the consumers, whose energy consumption process is coupled to their computational service and is represented by a stochastic service process. Thus in this model the EPs constitute the "ordinary customers" of the queueing network, the STs are the queues, the external arrivals of EPs are the energy produced by different sources of energy, and the G− network's "triggers" are the data flows regarding requests made by consumers or by STs whose energy buffers are emptied by the Cloud servers, and are replenished by the renewable or other energy sources.

The system we consider has a set G of energy sources each with an energy generation rate $g(i,t)$ in EP/sec at time t, for $i \in E$, where $g(i,t) \leq G_M(i)$ which is its maximum generation rate. The energy sources are either renewable, in which case $i \in R$, or they are conventional in which case $i \in C$.

The system has a set of S of energy storage centres (ST) each with finite storage capacity $K(j)$, $j \in S$. Each storage centre has an energy conversion efficiency $0 < e_j \leq 1$ at its input so that on average the arrival of $B'(j)$ energy packets to ST j results in the storage of $B(j) = e_j.B'(j)$ EPs. Furthermore, in addition to its maximum energy storage capacity, it will also have a maximum rate $\theta(j)$ at which it can store energy. It also has an energy loss rate which is β_j per unit time so that if storage is not replenished, the $B(j)$ EPs that S(j) contains will be depleted on average after a time $B(j)/\beta_j$. In addition, a ST will have a finite maximum rate $D(j)$ at which it can deliver energy. Let $d(j,t)$ be the instantaneous delivery rate of the ST at time t and we will have $d(j,t) \leq D(j)$. In fact, when energy is converted back from stored energy to a flow that is dispatched to another storage or consumer centre, there will again be an internal conversion loss, but we will include it within the parameter $e(j)$.

We also have a set of C Cloud computing centres (CC) which are the main energy consumers. The $c--th$ centre $C(c)$ has an a consumption rate of $m(c,t)$ in EPs/sec at time t. Some of these CCs may also have the ability to store energy locally. A Data Centre with its uninterruptible power supply (UPS) is an example of subsystem composed of a CC and a ST.

These centres are interconnected by an energy distribution or transport network (EDN) represented by a graph, so that link (u, v) of the directed graph represents a power line that has an energy transport capacity $C(u, v)$ which is the maximum amount of power that can be transferred instantaneously from node u to node v. In addition the link will have an efficiency $0 < c_{uv} \leq 1$ which is the fraction of energy introduced into the link that actually reaches the destination. The nodes of the EDN may be production nodes, consumption, storage nodes, or they may also be *transduction nodes* which can have many inputs and outputs. A transduction node u does not generate or store energy but dispatches it from one or more nodes to one or more other nodes; it has a transduction power capacity $T(u)$ so that for any successor node v we have $\sum_v C(u, v) \geq T(u)$ and for any predecessor v we have $\sum_v C(v, u) \leq T(u)$. Thus the incoming link capacities to a transduction node cannot exceed its own capacity, while the transduction node's outgoing links need to have a total capacity that exceeds its own capacity. The transduction node u will also have an efficiency $0 < t(u) \leq 1$ so that a fraction $t(u)$ of the power that it receives is wasted.

Each CC will send its energy requests to some Smart Dispatching Centre (SDC). SDCs are facilities that are interconnected to system components via a computer– communications network. Each SDC keeps track of the energy needs and requests in an area and assigns flows from the STs and EGs to the CCs. The SDCs also send requests to the EGs so that they may replenish the STs.

The SDCs' role is to satisfy the requests of the CCs and to make sure that STs have a standby capacity to meet unexpected needs. The SDCs will typically use pricing policies help the STs replenish their power at the best price, and also when energy from photovoltaic, wind or from other renewable sources is more readily available. The SDCs also attempt to maintain a flow of EPs across the EDN which is as low as possible so that energy traffic peaks are avoided and the EDN avoids saturation. Indeed one of the overarching objectives of the energy packet system is to be able to operate reliably with the lowest overall load being carried by the EDN.

Since the system as a whole depends on constant sensing, monitoring, communicaton and decision, the computer servers and network equipment will also constantly consume energy and this needs to be included in the model. In particular we will assume that every unit in the system receives a constant flow of energy from some of the energy sources (e.g. generators) for this purpose. The assumption then is that each of the units in the system, whether it be a ST or a CC, is aconnected to the generators to receive a flow of energy independently of its own requests based on its consumption, to assure that the information processing and communication systems can operate in a non– stop manner. Of course the generators themselves will be similarly monitored and connected to power sources (possibly the local units) but this aspect is neglected in the model. The purpose of these continuously running power supply from the generators to the ST and CC is also to allow them to receive some flow of energy over and above what they may request in order to allow their systems to operate all of the time on standby.

3 G– Network Model of the EPN

Based on the previous presentation, a queueing network analysis of the system that we have described can be constructed based on a stochastic representation of energy production, storage and consumption. The only element that we will not include in the model are the transduction nodes, but these can be included in future studies. We represent by the probability π_{uv}, the fraction of energy leaving node u which is directed towards node v, while $p(u, v) = \pi(u, v)c(u, v)$ is the fraction of the energy that leaves node u and *actually arrives* at node v. Furthermore, we also represent the effect of the SDC by its effect on the manner in which energy is actually being requested and dispatched, so that $q(v, u)$ will be the probability that when node $v \in S \cup C$ consumes or dispatches energy to some other node, then it requests energy from some other node $u \in G \cup S$.

The time behaviour of the generators will be represented by Poisson flows of rate γ_g for $g \in G$, while the time dependent energy consumption of consumers will be represented by a rate parameter μ_k for $k \in C$. The leakage of loss rate of a storage unit will be represented by an exponential distribution of parameter β_j, while the storage units' instantaneous energy output cannot exceed some given rate δ_j for $j \in S$. Each of the generators g will also attempt to provide a flow ϕ_g of standby energy to the storage and consumer centres, as indicated above, although the net standby energy flow it supplies will be $\phi_g \rho_g$ if ρ_g is the probability that the generator actually is able to supply energy, and will obviously be strictly less than one in the case of renewable and intermittent sources. As a result of these definitions, we can write equations that represent the equilibrium behaviour of the whole system using G– network theory as follows, where the main approximation concerns the formula that we will use to represent the finite capacity of the storage units. Since the storage units have a maximum energy delivery rate γ_j we will also define the parameter D_j as the actual energy delivery rate of the ST j:

$$D_j = max[\delta_j, \sum_{k \in C} \rho_k \mu_k q_{kj} \pi_{jk} + \sum_{i \in S} \rho_i \delta_i q_{ij} \pi_{ji}] \qquad (1)$$

Applying G-network theory [3,4] we have the following system of non-linear relations which describe the flow of energy into the storage centres and the consuming centres into which energy arrives on demand:

$$\Lambda_j = \sum_{g \in G} \phi_g \rho_g s_{gj} c_{gj} + \rho_j D_j [\sum_{g \in G} \rho_g q_{ji} p_{ij} \qquad (2)$$
$$+ \sum_{i \in S} \rho_i q_{ji} p_{ij}], \ j \in S$$

$$\Lambda_k = \sum_{g \in G} \phi_g \rho_g s_{gk} c_{gk} + \rho_k \mu_k [\sum_{g \in G} \rho_g q_{kg} p_{gk} \qquad (3)$$

$$+ \sum_{j \in S} \rho_j q_{kj} p_{jk}], \ k \in C$$

where s_{gj} is the fraction of standby energy that a generator g supplies to the ST or CC j, while $c(g, j)$ is the fraction of this energy that actually gets to the destination. The unknown terms $\rho_i, \ i \in S \cup G$ represent the probability that the corresponding storage centres and generators have energy to provide, while $\rho_k, \ k \in C$ represents the probability that a consumer centre is busy consuming energy. These quantities are expressed as functions of the $\Lambda_i, \ i \in S \cup C$:

$$r_k = \frac{\Lambda_k}{\mu_k}, \ k \in C, \qquad (4)$$

$$r_j = \frac{\Lambda_j}{\beta_j + D_j}, \ j \in S, \qquad (5)$$

and for any unit that has finite capacity, such as a storage centre that can at most store K_j EPs, we have:

$$\rho_j = r_j \frac{1 - r_j^{K_j}}{1 - r_j^{K_j+1}} \qquad (6)$$

Note that (6) is an approximation that is exact when we deal with a single queue with Poisson arrivals and exponential service times. We use it as an approximation in the networked systems that we are considering. Furthermore the case with $K_j = 1$ also will apply to a data centre that is limited to storing the energy that it is instantaneously consuming. For example during a machine cycle, all the energy that will be used in the cycle is already stored in the circuits of the computer which is drawing energy from its power supply as it computes successive cycles. When $K_j = 1$ we have

$$\rho_j = \frac{r_j}{r_j + 1} \qquad (7)$$

and

$$r_j = \frac{\rho_j}{1 - \rho_j} \qquad (8)$$

Since γ_g represents the nominal power generating rate of generator g at some given time, where g can be either a non– renewable or renewable source of energy, it will of course depend on the period of time being considered. For a non– renewable source will generally be possible to vary γ_g within a lower and upper bound, though changes may have to be carried out slowly over time. For a renewable sourge γ_g will generally depend on the instantaneous conditions (e.g. wind speeds, levels of ambient lighting for photovoltaic sources) but there are also ways to limit its value if it is deemed to be too high. We can therefore obtain the probability that the generator is able to satisfy all of its demands, given by:

$$\rho_g = min[1, \frac{\gamma_g}{\phi_g + \sum_{k \in C} \rho_k \mu_k q_{kg} \pi_{gk} + \sum_{j \in S} \rho_j \delta_j q_{jg} \pi_{gj}}] \tag{9}$$

If the CCs also have unlimited local storage capacity, then from (4) we have:

$$\rho_k = \frac{\sum_{g \in G} \phi_g \mu_k \rho_g s_{gk} c_{gk}}{1 - \sum_{g \in G} \rho_g q_{kg} p_{gk} - \sum_{j \in S} \rho_j q_{kj} p_{jk}}, \quad k \in C \tag{10}$$

In the special case where the STs have infinite storage capacity $K_j = \infty$ so that $\rho_j = r_j$, and the STs' loss rate is negligible $\beta_j = 0$ we also have:

$$\rho_j = \frac{\sum_{g \in G} \frac{\phi_g}{D_j} \rho_g s_{gj} c_{gj}}{1 - \sum_{g \in G} \rho_g q_{ji} p_{ij} - \sum_{i \in S} \rho_i q_{ji} p_{ij}}, \quad j \in S \tag{11}$$

In order to have enough energy for all the needs of the system, these equations would have to satisfy the constraint:

$$\sum_{g \in G} \gamma_g \geq \sum_{g \in G} [\phi_g + \sum_{k \in C} \rho_k \mu_k q_{kg} p_{gk} + \sum_{j \in S} \rho_j \delta_j q_{jg} p_{gj}] \tag{12}$$

so that the total energy that reaches the storage centres and the consumers do meet the rate at which they make their demands. Better still, ideally we would also like all the consumers' energy needs to be satisfied:

$$\rho_k = 1, k \in C \tag{13}$$

4　A Special Class of EPNs

A special class of EPNs that has intuitive appeal would have the renewable energy sources feed an overwhelmingly large part of their energy production directly into storage units except for a small fraction that is sent to the data centres just to keep them awake in case the storage centres are down. This has the advantage of buffering the fluctuations of renewable energy sources, but will also lead to potentially higher storage and energy conversion losses. In this case after some analysis we can obtain results for two cases of interest.

Assume that the C data centres that are providing processing for the Cloud are all identical and have the same statistically identical computational load. Suppose that they all have unlimited local energy storage. Suppose also that the ensemble of S storage centres are identical with unlimited capacity. and let l be the fraction of power that is lost in transit through the energy transmission network. Assume also that the storage centres do not have any conversion losses. After some calculations we can show from the preceding analysis that the probability that any one Cloud service centre has enough energy to meet its computational load is given by:

$$\rho_c = \frac{\gamma_g G(1-l)}{C \mu_c}[(1-s) + s(1-l)] \tag{14}$$

where s is the fraction of energy that is sent directly to the storage centres.

On the other hand, under similar conditions, if the Cloud service centres do not have the ability to store energy locally so that there may be energy wastage from the energy they receive but do not absolutely need to run their computations (for instance, they may use the energy to cool down their equipment at a higher rate or to keep their machines running at a higher rate than is needed, then with the same level of energy production γ_g the probability that one of these data centres is unable to meet its computational load will be:

$$\rho_c^* = \frac{\frac{G}{C}\gamma_g(1-l)[(1-s)+sl]}{\mu_c + \frac{G}{C}\gamma_g(1-l)[(1-s)+sl]} \tag{15}$$

so that quite obviously $\rho_c > \rho_c^*$. This simply states that in addition to shared energy storage it would also be useful to have local energy storage at the data centres. However this analysis has not included the effect of energy loss at the storage units, nor of loss due to reconversion, although this can be included in the loss factor l. Thus the results will be mitigated when one considers these additional losses in the storage units.

5   Conclusions

This paper discusses a novel system that we call an *Energy Packet Network* to store renewable or cheap electric energy and deliver it on demand to computational systems in the Cloud. This approach can have great value whenever scarce or valuable sources of energy must be shared by multiple computational units whose peak to average power consumption ratio is high so that storage can smooth the intermittent supply and the intermittent demand. A method for the analysis of such systems based on queueing networks is suggested. The analysis is applied to a special case, indicating that it would be advantageous to have local energy storage at the data centres as well as energy storage units that are shared among many different data centres. In future work we plan to show how SEDCs can use distributed adaptive schemes such as those developed for smart packet network routing [10,19] so that when a request for cloud computing arises, the SEDCs will select the computational resource which can do this at lowest energy and/or economic cost, and dynamically provision the energy dynamically.

Acknowledgements. The author gratefully acknowledges the support for this research from the Fit4Green European Union FP7 Project co- funded under ICT Theme: FP7- ICT- 2009- 4.

References

1. Bunn, D.W., Farmer, E. (eds.): Comparative Models for Electric Load Forecasting. John Wiley & Sons (1985)
2. Gelenbe, E., Stafylopatis, A.: Global behaviour of homogeneous random neural systems. Applied Mathematical Modelling 15(10), 534–541 (1991)

3. Atalay, V., Gelenbe, E.: Parallel algorithm for colour texture generation using the random neural network model. IJPRAI 6(2&3), 437–446 (1992)
4. Gelenbe, E.: The first decade of G-networks. European Journal of Operational Research 126(2), 231–232 (2000)
5. Ramanathan, R., Engle, R., Granger, C.W., Vahid-Araghi, F., Brace, C.: Short-run forecasts of electricity loads and peaks. International Journal of Forecasting 13(2), 161–174 (1997)
6. Gelenbe, E., Labed, A.: G-networks with multiple classes of signals and positive customers. European Journal of Operations Research 108(2), 293–305 (1998)
7. Winkler, G., Meisenbach, C., Hable, M., Meier, P.: Intelligent energy management of electrical power systems with distributed feeding on the basis of forecasts of demand and generation. In: CIRED 2001 (2001)
8. Zack, D.J.: Overview of wind energy generation forecasting. Tech. Rep. TrueWind Solutions, LLC (2003)
9. Gelenbe, E., Fourneau, J.-M.: G-Networks with resets. Performance Evaluation 49, 179–192 (2002)
10. Gelenbe, E.: Cognitive Packet Network. U.S. Patent No. 6804201 B1 (October 12, 2004)
11. Taylor, J.W.: Density forecasting for the efficient balancing of the generation and consumption of electricity. International Journal of Forecasting 22(4), 707–724 (2006)
12. Gelenbe, E., Loukas, G.: A self-aware approach to denial of service defence. Computer Networks 51(5), 1299–1314 (2007)
13. Cancelo, J.R., Espasa, A., Graffe, R.: Forecasting the electricity load from one day to one week ahead for the spanish system operator. International Journal of Forecasting 24(2), 588–602
14. Black, M., Strbac, G.: Value of bulk energy storage for managing wind power fluctuations. IEEE Transactions on Energy Conversion 22(1), 197–205 (2007)
15. Infield, D., Short, J., Home, C., Freris, L.: Potential for domestic dynamic demand-side management in the UK. In: IEEE Power Engineering Society General Meeting, pp. 1–6 (June 2007)
16. Dordonnat, V., Koopman, S., Ooms, M., Dessertaine, A., Collet, J.: An hourly periodic state space model for modelling French national electricity load. International Journal of Forecasting 24(4), 566–587 (2008)
17. Sanchez, I.: Adaptive combination of forecasts with application to wind energy. International Journal of Forecasting 24(4), 679–693 (2008)
18. Manwell, J., McGowan, J., Rogers, A.: Wind Energy Explained: Theory, Design and Application. Wiley (2009)
19. Gelenbe, E.: Steps toward self-aware networks. Comm. ACM 52(7), 66–75 (2009)
20. Berl, A., Gelenbe, E., di Girolamo, M., Giuliani, G., de Meer, H., Dang, M.-Q., Pentikousis, K.: Energy- efficient Cloud Computing. The Computer Journal 53(7), 1045–1051 (2010), doi:10.1093/comjnl/bxp080
21. Sakellari, G., Gelenbe, E.: Demonstrating cognitive packet network resilience to worm attacks. In: Proc. ACM Conference on Computer and Communications Security, pp. 636–638 (2010)
22. The MeRegio Project, http://www.meregio.de/en/ (2011), Center for Renewable Energy Sources, http://www.cres.gr/
23. Berthold, H., Boehm, M., Dannecker, L., Rumph, F.-J., Pedersen, T.B., Nychtis, C., Frey, H., Marinsek, Z., Filipic, B., Tselepis, S.: Exploiting renewables by request– based balancing of energy demand and supply. In: Proc. 11th IAEE European Conference (2010)

24. MIRACLE Project 2010. MIRACLE Project Website. MIRACLE Project (2010),
    http://www.miracle--project.eu
25. Nationalgrid UK 2010. Metered half-hourly electricity demands. Nationalgrid UK
    (2010), http://www.nationalgrid.com/uk/Electricity/Data/Demand+Data/
26. NREL 2010. Wind Integration Datasets. NREL (2010),
    http://www.nrel.gov/wind/integrationdatasets/
27. Gelenbe, E., Morfopoulou, C.: Routing and G-Networks to Optimise Energy and
    Quality of Service in Packet Networks. In: Hatziargyriou, N., Dimeas, A., Tomtsi,
    T., Weidlich, A. (eds.) E-Energy 2010. LNICST, vol. 54, pp. 163–173. Springer,
    Heidelberg (2011)
28. Gelenbe, E., Mahmoodi, T.: Energy-Aware Routing Protocol in the Cognitive
    Packet Network. In: International Conference on Smart Grids, Green Communica-
    tions, and IT Energy–aware Technologies (Energy 2011), Venice, Italy, May 22-27
    (2011) ISBN: 978-1-61208-006-2
29. Gelenbe, E., Morfopoulou, C.: A framework for energy aware routing in packet
    networks. The Computer Journal 54(6), 850–859 (2011)
30. Dinorwig power station. First Hydro Company,
    http://www.fhc.co.uk/dinorwig.html
31. Bitar, E., Rajagopal, R., Khargonekar, P., Poolla, K.: The role of co-located storage
    for wind power producers in conventional electricity markets. In: Proc. American
    Control Conference (ACC), pp. 3886–3891 (July 2011)
32. Chandy, K., Low, S., Topcu, U., Xu, H.: A simple optimal power flow model with
    energy storage. In: 49th IEEE Conference on Decision and Control (CDC), pp.
    1051–1057 (December 2010)
33. Gayme, D., Topcu, U.: Optimal power flow with distributed energy storage dy-
    namics. In: Proc. American Control Conference (2011)
34. Grünewald, P., Cockerill, T., Contestabile, M., Pearson, P.: The role of large scale
    storage in a GB low carbon energy future: Issues and policy challenges. Energy
    Policy 39(9), 4807–4815 (2011)
35. Gelenbe, E.: Energy Packet Networks: Smart Electricity Storage to Meet Surges
    in Demand, Keynote Talk. In: SimuTools 2012, Desenzano, Italy (April 2012)
36. Naish, C., McCubbin, I., Edberg, O., Harfoot, M.: Outlook of energy storage tech-
    nologies. Technical Report
37. Oh, H.: Optimal planning to Include Storage Devices in Power Systems. IEEE
    Transactions on Power Systems 26(3), 1118–1128 (2011)
38. Sinden, G.: Characteristics of the UK wind resource: long– term patterns and
    relationship to electricity demand. Energy Policy 35(1), 112–127 (2007)
39. Su, H.-I., Gamal, A.E.: Modeling and analysis of the role of fast-response energy
    storage in the smart grid. In: Proceedings of the Forty-Ninth Annual Allerton
    Conference on Communication, Control, and Computing. University of Illinois at
    Urbana-Champaign (September 2011)
40. Su, H.-I., Gamal, A.E.: Modeling and analysis of the role of fast– response energy
    storage in the smart grid. CoRR, abs/1109.3841 (2011)
41. Financial Times, p. 2 (August 29, 2011)
42. Wade, N., Taylor, P., Lang, P., Jones, P.: Evaluating the benefits of an electrical
    energy storage system in a future smart grid. Energy Policy 38(11), 7180–7188
    (2010)
43. Gelenbe, E.: Energy Packet Networks: Smart energy storage to meet surges in
    demand. In: Proc. 5th International ICST Conference on Simulation Tools and
    Techniques, Simutools 2012, Desenzano, Italy, March 19-23 (2012)

Author Index